THE QUANTITY SURVEYOR'S BIBLE

Ian Carroll

Oakamoor Publishing

Published in 2019 by Oakamoor Publishing, an imprint of Bennion Kearny Limited.

ISBN: 978-1-910773-69-7

Published by Oakamoor Publishing, Bennion Kearny Limited
6 Woodside
Churnet View Road
Oakamoor
Staffordshire
ST10 3AE

www.BennionKearny.com

TABLE OF CONTENTS

PROLOGUE

Welcome to the new version of The Quantity Surveyor's Bible. It's only three years since the book was first published, so why the need for a new edition? Well, the first edition was very concise. It was written, almost as an angry letter, in response to an employer who wanted me to do my job a certain way. It was very much my way of saying, 'No, this is how to be a good QS,' based on my own 20 years of experience in the field.

Once I had calmed down and had worked off my head of steam, I knew I would have more to say, and in a much more measured way. So, I have expanded on certain themes in the book, and added several others, such as the case studies. I've also been out in the freelance world in the interim, and gained experience in sectors I'd never previously had exposure to, such as new-build housing. I thought it was important to add that to this edition of the book.

And then I thought we'd also benefit by having a more attractive cover! One thing that has given me great pleasure since the book first came out is the many great reviews we've received on Amazon.

I read every one, and I am also constantly checking for new ones. If you do read the book, I'd be truly grateful if you could follow it up with a review – good or bad – and I can promise you that it will be read personally by me, your author and guide for this book.

Originally, the sub-title for The Quantity Surveyor's Bible was 'Adventures in the Construction Industry.' I hope that I can help you to have some adventures of your own. Thanks for reading.

Ian Carroll

INTRODUCTION

What is a Quantity Surveyor? Is he a construction professional? Is he the accountant of the construction industry? Is he just someone who counts bricks? Is he the most boring man on the planet?

Well, the first thing to say is that the QS doesn't have to be a 'he'. It could equally be a 'she'. There are many female Quantity Surveyors out there, and many women entering the building industry as a whole. The industry needs more women, and needs more people in general, especially skilled workers as well as professional people. But, as I'm a 'he', I'll use 'he' in my examples, if that's okay.

So, to answer the question above. What is a Quantity Surveyor?

Is he a construction professional? Yes.

Is he the accountant of the construction industry? Yes.

Is he just someone who counts bricks? No.

Is he the most boring man on the planet? Sometimes!

Let's not kid ourselves. We're not rock stars (although Nick McCabe of The Verve was a QS), we're not film stars (though screenwriter and television producer Phil Redmond was a QS), we're Quantity Surveyors; we work in the construction industry and our expertise lies in managing the costs of the construction process.

To be a QS, you need to be good at maths (or you may as well stop reading this now!), and you need to understand the construction process, i.e. how things are built. Once you have these two things in place, good maths and a knowledge of construction, you have everything you need to be a QS. How good a QS is up to you. And here's the thing, after more than 30 years in the construction industry, 20 of them as a Quantity Surveyor, I'm convinced that the job is fifty percent communication.

Can you talk to clients, to architects, to the contractor or subcontractors, to the lads (or girls) doing the actual work on site? If you can talk to all of these people, if you can understand, problem-solve, contribute, and motivate, then you have all the ingredients necessary to become a good QS.

I'm assuming you have an interest in becoming a Quantity Surveyor or some similar construction professional. I hope that the contents of this book will serve to inspire and enlighten you along the way.

THE CONSTRUCTION INDUSTRY

What is the Construction Industry? What does it do? Who does it employ? Who does it affect? Why is it important? Now this book is not going to be an overly-academic piece of work. Its aim is to be informative, but the last thing it's meant to be is dull. All answers are my own, based on my own experience, and I hope you'll find them useful, insightful, and engaging.

Hopefully, the content will inspire you to pursue a career in construction, maybe as a Quantity Surveyor or similar professional, and fully prepare you for the journey that lies ahead.

We will be talking about the built environment. This is the world of houses, schools, factories, stadiums, cinemas, hotels, shopping malls, towns, cities, apartment blocks, and skyscrapers, as well as the roads, railways, and canals that string them all together.

So, what exactly is the construction industry?

One of mankind's most basic requirements is the need for shelter. I'm sure we all built dens as children. Watch Bear Grylls on one of his survival adventures and – before he even goes out to look for food – he'll build himself a place to stay. From this humble abode, he can start to think about keeping warm and dry. There is also the need for protection and security. A man's home is his castle, as the old saying goes.

The housing market is just one aspect of the construction industry, but it's a fundamental part. We are millions (and billions) of people holed up in our dens. Once we're all safe in our homes, we can start to feel secure and then we might start to venture out in search of food and water. So, we need shops. Back in our homes, we've found a mate, and we start to rear children. We want to educate them; we build schools. Now the kids need shoes; we need shops, and jobs to pay for the clothes on our backs. We build factories and offices. We need cars to get to our jobs or to take the kids to school; so we build roads, bridges, and tunnels, and giant industrial plants to keep those cars rolling off the production line. All that hard work is making us ill. We need hospitals. Now we're well again, and we have a little spare cash. We want entertainment. We build theatres, cinemas, sports stadiums. And on and on it goes.

The built environment is all of these buildings; built by someone. Usually, a team of people, working hand in hand. Behind that ant-like endeavour, there's usually a plan drawn up by an architect, and some thought has gone into a programme of work so that everything happens in sequence. You can't put the roof on until you've got something in place to support it, like walls. So you get your brickies in first, and the roof can go on afterwards. And, unless you're an Arab sheik or a Russian billionaire, you're probably going to be conscious of the costs. It's a good idea, before you start, to know if you can afford to build whatever it is that you're planning to build. Otherwise, you might be facing financial ruin. No one wants to sign a blank check. So, someone has to take that architect's drawing – before a spade has even been stuck in the ground – and figure out how much this pretty picture is going to cost to build. That someone could just be you, in your capacity as a Quantity Surveyor.

Where do you live? Is there a building nearby that you like? Grab a pen and a piece of paper. Imagine when it was just a bare patch of earth. What happened next? Try and make a list. Machinery came in and dug a hole. Concrete was poured. Slowly, the building began to rise out of the ground. There were walls, windows, (eventually) a roof. Inside, teams of plasterers, painters, joiners, electricians, and plumbers were all hard at work. And somewhere, there was someone watching over the whole thing, making sure that everyone was sticking to the plan and to the budget. And that someone might just be you.

WHAT IS A QUANTITY SURVEYOR?

You go to a party, or the pub, or you meet someone somewhere, and they ask what you do for a living. In my case, I reply, "I'm a Quantity Surveyor." 'What's that?' is often the response that I get. So, what exactly does a Quantity Surveyor do?

There are many different types of QS roles, and many different types of Quantity Surveyors. Some are what I call accountant-type QS's. They are concerned purely with numbers, and they rarely look up to see whatever it is that they're building. It's all just numbers to them. Some of these people are very good QS's, and many of them are miles better than me. They know all of the current legislation, and they stick religiously to measurement techniques that are enshrined in such earnest textbooks as SMM7 or the New Rules of Measurement prepared by the RICS. They stick to the rigid path and provide a solid audit trail that befits their accountant-like status. That's one kind of QS.

For me, being a QS is really about five different things. It is about estimating costs for feasibility or tendering purposes. It is about the procurement of materials and labour to carry out the work. It is about valuing work done in order to apply for payments. It is about negotiating with clients or subcontractors. And, it is about project management in order to add value throughout the lifespan of the build process. These, as I see it, are the five key areas that make up the role of a QS.

It is also about communicating in order to keep the project on track, and avoiding conflict which might lead to contractual claims (therefore an understanding of the law of contract and the ability to communicate effectively is a must).

Now the construction industry is a giant beast with many different facets. It is no easy thing to say 'this is what a QS does' because you could case-study a hundred different QS's, and each of them would be doing a different thing, or even doing the same thing but in a completely different way. That said, we can talk about some of the things that a Quantity Surveyor is likely to be doing. We'll look at several scenarios and, the likelihood is, if you end up as a QS, you're probably going to be working in a similar environment to the one described, and doing a job that bears a close resemblance to the one we're talking about. So, what might a QS be doing; in what environment might he or she be doing it; and who might he or she be likely to be doing it for?

Assuming you are entering the world of QS'ing after a degree or diploma of study, the trainee / graduate QS will usually be faced with a fork in the road, to either enter private practice or become a contractor's QS.

Private practice is also known as 'client-side QS'ing'. The client being the person who will ultimately own the building. These positions probably account for between ten and fifteen percent of all of the available QS roles.

If you choose client-side private QS'ing, you will probably find yourself working in a nice office – wearing a shirt and tie – surrounded by admin staff, managers, and possibly designers, architects, and building surveyors. You will be employed by the client, the person or organisation who will ultimately own and use the building being constructed. You are there to ensure that they get good value for money and to make sure that the contractor – the person building the thing – complies with all of their contractual obligations. I like to think of it like this. As a client-side QS, once you have made your contribution by making an early estimate of the likely costs, and once a contractor has been appointed to construct the actual building, you are there to hold the bag of cash (not literally) that will fund the project and from which the contractor will be paid. Needless to say, you must exercise due diligence throughout, ensure accountability, and protect your client's interests from a financial point of view.

The majority of graduates, however, will enter what is called 'contracting' (i.e. working for the people who actually build whatever it is that is being built). These could be house-builders, or builders of industrial buildings, schools, hospitals and the like. Likewise, QS's could be employed within large civil engineering firms constructing roads, bridges, motorways, tunnels, dams, stadiums, and airports – the huge infrastructure that makes up a nation.

If you go into contracting, you may be site-based (though not always), and you will be involved more heavily in the actual construction process. You may have numerous subcontractors fulfilling certain aspects of the construction, and they will be dependent on you to award them work and to measure their applications for payment to ensure that you retain a winning margin for your employer once the project reaches completion. Being site-based, or at least closer to the actual build, you will find the language a lot coarser. You are, after all, dealing not with councils or clients with money (with presumably some sense of sensibility to match), instead, you will be at the coal-face, dealing with working people who will speak plainly and directly and who will expect clear answers (and proper payment!) as they go about their business at the practical end of the construction process.

Of the two QS environments, client-side QS'ing is usually a nicer, cleaner, more sophisticated world. Your office could be mistaken for an insurance company or some other such sterile space, though they are great places to work and I have no problem with them at all. If, however, you want to get your hands dirty – sometimes literally – and you want to feel the wind on your face, and you want to see things being built, and you actually like a bit of hustle, bustle, and tussle, then contracting may be the way to go. And, if you do decide on the latter, at least you know that 85% of the available job opportunities for a QS will be on that side of the fence.

That said, you can fluctuate (as I have) throughout your career, and that's no bad thing either. Try one, and then try the other. I've been the man paying the money, and I've been the man asking for the money. I believe it makes you a more rounded QS. Whichever area of QS'ing you choose to go into, remember that – at the end of the day – the one thing that doesn't change is that you're a Quantity Surveyor and a construction professional.

THE ACCOUNTANT OF THE CONSTRUCTION INDUSTRY

The QS is often called the accountant of the construction industry. I've even heard the expression the economist of the construction industry. Why's that?

Well, the first thing to say is that I believe Quantity Surveying to be a whole lot more interesting than accounting, or economics for that matter. Accountancy is something done on the page. It is concerned with numbers, with little thought as to what those numbers actually represent. At least, that is how it seems to me. An actual accountant may beg to differ.

I think Quantity Surveying is about numbers that relate to real bricks and mortar, real projects, tangible things, even something as significant as a skyscraper. These aren't things lying flat on the page. These are objects you can look at, live in, touch, and walk around in. You can stare at them in wonder. It's a different thing, as I see it, to pure accountancy. It's a lot more exciting. A lot more real. I think it was Einstein who said 'Not everything that counts can be counted, and not everything that can be counted, counts.'

So why do they call Quantity Surveyors the accountants of the construction industry? Simple. Because our job relates primarily to costs. Imagine if a client wants to build a new house, or a local authority wants to build a new hospital. First of all, they might think about their budget. How much money have they got to spend? Say our client is a Premier League footballer. He's got a million pounds to spend on a new home. He's already bought a plot of land. Now, all he needs to know is what kind of house he can get for his budget.

He engages an architect. In my view, architects are great. They are clever, knowledgeable people – by and large – and they are creative as well as technical. They're usually a little quirky while coming across as all professional. A great bunch, on the whole, but I digress. So, the architect asks what the client wants. Four bedrooms, three bathrooms, a large kitchen / dining room, a big living room, a swimming pool, and a double garage is the outline he gets given.

The architect gets to work. He draws a house that expresses something of the architect's own taste and hopefully fulfils the client's brief. That's all well and good, but how much is the house going to cost? This is where we need a cost consultant, usually a Quantity Surveyor, to answer the million-dollar question. Now, guess what… the truth is that – at this stage – no-one knows for certain what the *actual* cost is going to be. All the QS can do is produce an estimate of the cost. He (or she) is going to say 'I reckon that house is going to cost you a million quid' or whatever it is that his estimate comes out to. The reality is that, until you find a builder who is going to build the house for a fixed fee, usually having gone through the tender process in competition with two or three other builders, then the cost of the project remains speculative. Until you actually shake hands on the deal, to build such and such a house at such and such a cost, then the estimate is just an educated guess by a construction professional (e.g. you) using the best of his or her experience and expertise.

It's called an 'estimate' for a reason, because that's what it is. It's not an 'exactimate'. Such a phrase does not exist. Once you give a client an estimate, e.g. 'I reckon that house of yours is going to cost a million pounds to build', then the truth of how well you've done will *only* be revealed once the actual tenders are returned. If you get three quotes back from different builders, with the lowest at £950,000, the middle one at £1,000,000, and the highest at £1,050,000, then you know that your 'guesstimate' was pretty accurate. If, however, the three tenders come back at £1.2 million, £1.4 million, and £1.6 million then you've got a lot of explaining to do. Either you're having a bad day at the office or you're in the wrong job. Believe me, get too many of them wrong, at significant margins, and you won't have a job for long. Not as a QS, anyway.

So that's the first instance of a Quantity Surveyor coming into contact with numbers, at the feasibility stage. There are many more examples, right through the lifespan of a construction project, from tender stage, to your own cash-flow forecasts, calculating your profit margin, applying for interim payments, and ultimately settling the Final Account; so, the term 'accountant of the construction industry' is well-deserved.

As for being the economist of the construction industry (a term used far less frequently), I don't happen to believe that either. When I studied QS'ing at university, one of our first-year subjects was economics, but I don't remember a whole lot about it. I found it incredibly boring (which QS'ing is not!). Of course, it wasn't helped by the fact that it was the last lecture of the day, so the interest-level of the class was at a record-low by then.

Still, apart from understanding that if the economy of a country is booming, a lot more building will be going on; and if the economy is in a mess, then nothing much of anything will be happening, then I don't really think we need to consider if a QS is the 'economist of the construction industry'. I think 'accountant' is more accurate but, as you know, I don't accept that one either. We're Quantity Surveyors. That's all that needs to be said.

I did accountancy in school. I remember there was a lot of stuff about double-entry book-keeping and balancing the books. As a Quantity Surveyor, one of the things you'll often have to do at the end of a project is a Cost-Value Reconciliation exercise. This is known as a CVR.

This involves calculating all of the costs racked up against the job: the labour, materials, and plant expenditure, as well as any supervision, overheads, etc, apportioned to the job. This will tell you how much it actually cost to carry out the work. Then, you look at the value, which is the money actually earned for doing the job, meaning the money that your client paid you for carrying out the work. Then, you reconcile the two. If it cost you forty grand to do the job, and you charged the client fifty grand for it, then it doesn't take a genius to work out that you made a ten grand profit. That's your CVR.

I'll qualify that last statement, because profit is a misleading term in this instance. That ten grand we've just made also has to cover our overheads. If you work in an office, and your company has an office manager, and a receptionist, and computers, and telephones, and company vehicles, and staff, then probably fifty percent of what we called profit in the previous example will be eaten up by overheads. Your profit will, therefore, be the other fifty percent. So five grand overheads and five grand profit. Overheads and profit go hand in hand in any reconciliation exercise.

We'll cover estimating and tendering in due course. For now, imagine that when you came to put a price in for the job, you calculated what your labour costs might be. You calculated what your material costs might be. You put an allowance in for any plant, e.g. scaffold, skips, etc. You worked out the total cost of all these elements,(you worked out what the job was going to cost *you* to do), and then you put a percentage mark-up (say thirty percent) to cover *your* overheads and profit. That's your tendered sum; the cost of the job plus a margin for overheads and profit.

The Quantity Surveyor can be a cost consultant, a commercial manager (you'll be referred to as part of the commercial team anyway), or a client's representative. The latter phrase is a relatively new term that seems to be growing in popularity. As a client's representative, you'll be expected to represent the client's interests, and all of your QS skills – those of communication, problem-solving, negotiation, and project management, as well as cost consultancy – will all be called into play.

Have you got the skills to succeed? I hope so. But, like I've said before, the one thing you won't be is an accountant, not in the usual sense of the word. You'll be a Quantity Surveyor, and I hope that you'll be a good one.

WHAT QUALIFICATIONS DO YOU NEED TO BE A Q.S.?

I have a BTEC HND in Quantity Surveying and an NVQ Level 3 in Surveying Support. I also have 20 years Quantity Surveying experience, and a Master's Degree in Writing. The more educated amongst you may question my qualifications to write a book like this. I would only answer, as stated earlier, that not everything that counts can be counted. This is my QS story. You can always write your own.

I was 25 when I decided to embark on a career as a QS (I had already spent five years working in the building industry at that point, working as a labourer, and then as a roofer's mate, then as a roofer, but I wasn't very good working with my hands). One day, frustrated with my lot, I asked a colleague on site who the bloke wearing the tie was, the fella walking around the building site talking to various trades-people, the foreman, etc, and heading in and out of the site office. I was told that he was the Quantity Surveyor. What's one of them, I wanted to know. I was told that you had to be good with numbers and you had to understand construction. I thought, I can do that, so I set my mind to the task.

I spent a year in college doing a BTEC Foundation course in Construction which served as my access route into university. Then, as I was already in my mid-twenties, I decided to go for an HND rather than a full degree. It was two years full-time at university. With the foundation course, it meant three years out of my working life, three years without a regular income. As my real dream was to be a writer someday, I only needed to reach technical, not professional level. I thought, it might be 20 years before I got my break. What did I want to do for those 20 years? Did I want to pull pints? (Apologies to all you bar-staff out there!) No. I wanted to have a decent day job, doing something that I enjoyed, and earning a good salary. It turned out to be one of the better decisions that I've made in my life, and I don't regret it at all.

Eventually, as I went from graduate QS to assistant QS, I undertook an NVQ in Surveying Support, covering the technical aspects of construction operations, as part of a work-based study course while working in a private QS practice. Finally, I racked up enough experience that the 'assistant' tag just fell away and I found that I had reached my goal. Since that day, 20 years ago, I have worked as a QS for small, medium, and large contractors, in consultancy, as a freelance QS, and as a Claims Consultant. It's given me an all-round view of the industry, and one that I hope to share with you within the pages of this book. I hope that you'll find it useful, and that it will inspire you to go on your own QS adventure in the construction industry. I believe that it's worth it.

Most people studying quantity surveying do so through a Bachelor of Science degree. Some follow this up with a Master's degree, and many more follow the route of the RICS, the Royal Institute of Chartered Surveyors, to become Chartered Surveyors.

Some of you may be coming to the profession late, maybe as a career change. I know a lad who has a Master's degree in Curating and a degree in Fine Art. He is now on a one-year QS conversion course. The 32 graduates from the previous year are ALL now employed as QS's. Already, employment agencies are calling me up asking when my friend will graduate. Quantity Surveyors are very much in demand, even those without years of experience or membership of a professional body behind them.

Personally, apart from once sitting the Tech RICS interview and not succeeding, I've managed to notch up a pretty decent CV and a career without ever joining the RICS. A former line-manager of mine, and Associate of the Private Practice where I worked, explained why he had chosen the RICS route. In the early nineties, during a major recession that fairly decimated the UK construction industry, when he was struggling to find employment, he decided that he'd give himself every opportunity in the future to be able to compete with the best candidates around. He was going to get all of the qualifications needed to prove to potential employers just how capable he was. He had a family to feed. He didn't want to see that job – if only one there was – go to someone else who appeared, on paper, more qualified.

I can't fault his logic, but many QS's succeed without the endorsement of the RICS. If you're able to do the job, hopefully you'll get ample opportunity to show it.

At the same time, I worked in an office where there was a surveyor who would never give you a straight answer. If asked a question, he would just say 'I'll have to look it up.' Well, as someone once said to me, 'we could all do that!'

In that same office, there was another surveyor who was old school, who had great practical knowledge, and who would answer every question and could advise you on just about anything. These two men both applied for membership to the RICS. Guess which one got accepted? Correct. The one who could never answer a question without looking it up.

So, as Groucho Marx once said, 'I wouldn't want to join any club that would have me as a member!' That said, I have to take my former manager's advice and, if you want to compete with the best, there's no harm in giving yourself every advantage with which to do so. It's your life, and your career, so you decide. For me, being a good QS is enough.

STARTING OUT

Once you have completed a course of study, you will hopefully find employment with a contractor or private practice and begin your quantity surveying career. You may find yourself with the title of either trainee or graduate QS. If you have already had a bit of experience in the role or in life in general, or have transferred from a similar discipline, you may earn the title of assistant QS. You're well on your way to being a fully-fledged Quantity Surveyor.

You will probably find yourself doing some of the donkey-work. This might involve taking off quantities from an architect's drawing, e.g. measuring the floors, walls, or ceilings, to help make up a bill of quantities, which forms an essential part of the tender documents that building work is priced on. The bill of quantities, often abbreviated to BOQ, contains a breakdown of the individual components that make up that particular building. It will state, in metres squared or linear metres, how much brickwork there is, how much plastering, how many doors, windows, etc, for everything from the footings in the foundations to the ridge of the roof. You'll get some exposure to life on site to build up your construction knowledge, and you'll attend pre-start and progress meetings to get used to the communication side of things (meeting the client, architect, and other stakeholders in the project).

You probably won't be expected (or encouraged) to say too much at this point. Instead, this is your opportunity to listen and learn. One important thing to mention here is that, ultimately, we are employed in the construction industry. As important and interesting as this industry can be, it is populated by normal people, with buildings being ultimately constructed by your average person, e.g., bricklayers, plumbers, electricians, painters and decorators and the like. We're not rocket-scientists, and we're not saving lives. In other words, there's nothing being said or done that is outside an average person's understanding. So, keep calm, and don't feel the need to make a contribution just for the hell of it. You have permission, at this stage of your career, to sit back and observe. Everything learned will ultimately serve the surveyor that you're going to be, but you're not there yet, and nor will anyone expect you to be. Just relax, enjoy the ride, and soak up as much information as you can. Your time will come, believe me, and people always appreciate a calming influence at the table. No-one likes a bull in a china shop.

How long will your training last? When will you step up from trainee or graduate to assistant, and from there to fully-qualified QS? Well, the answer to that is up to you, and also up to your employer to some extent. The latter might not recognise your inherent skills, or your frustration at performing a role that you feel you have outgrown. In that case, you need to speak up for yourself. It's communication, remember. You're also displaying your skills of negotiation. You need that too. Ultimately, you're showing them that you have what it takes to be a QS. You're demonstrating the very attributes that you're asking them to recognise.

Eventually, as a result of getting a few years of experience under your belt, or some further qualifications or RICS membership, you'll lose the tag of trainee, graduate, or assistant, and you'll be a Quantity Surveyor. As long as you can do the job, you'll get there in the end. If you're a trainee at 20, you should be a QS by your mid-to-late twenties. If you're a graduate, with a degree or HND, at 21, I'd say you'd be an assistant at 23 and a QS at 25. Basically, unlike me, if you start early, you should be there by your mid-twenties. If you're a late starter like me (mid-twenties) then you should be there by the time you're 30. Feel free to start later, or change careers, and you'll still get there eventually. The important thing is, a winner sets their eye on the prize and focuses until they get there. If you want to be a QS, and you're good at maths and you understand construction, then you *will* get there. And all the better if you can communicate effectively, because (as mentioned previously) I believe that is fifty percent of the job.

Should you choose to go the route of the RICS, then you will have to face an interview and sign up for CPD – Continuing Professional Development – keeping abreast of changes in the industry (which we all do anyway) and being 'up' on the latest legislation. The RICS interview is known as the APC, which is the Assessment of Professional Competence. It's all in the name, really. Typical questions could be 'Outline the function of feasibility studies on projects', 'How do you value variations against a schedule of works', or 'What do you do if the lowest tender is also the lowest quality in terms of specification?' Remember, you'll be sitting in front of a panel of experts at this point, so both your knowledge and communication skills will be to the fore. I'll attempt my own answer to those questions at the end of this chapter.

Finally, in terms of qualifications, I'd just like to add one further thing. If you go to your local Yellow Pages and find the entries for Quantity Surveyors, you may see that a sizeable proportion of people listing themselves as QS's have certain credentials following their names. They may be BSc, or RICS, or FRICS, or MCIOB (Member of the Chartered Institute of Builders). Probably about half of them will also include LLb. This is a law degree. Why the close relationship between law and quantity surveying? Well, as part of any QS study course, you'll acquire a basic understanding of Common law and a more in-depth understanding of the Law of Contract. After all, buildings are built to a contract. The most basic form of contract is a simple handshake, but most construction projects require a little more than that.

When things go wrong (which we'll look at in more detail a little later), things can get 'contractual'. At that point, whether you're going to mediation, arbitration, adjudication, or even litigation (i.e. court), you'll need an understanding of contract law. Many QS's are now employed in the growing field of Claims Consultant, dealing with contractual claims. Maybe all of your jobs will run smoothly. Maybe all of your clients will be non-confrontational. But maybe not. Enjoy your module in Contract law. I know I did. And please pay attention in class. You may just need to call upon it more than you thought you would.

So, how would I answer the three RICS questions above? Well, I can't remember what the required word-count was, but I'll answer them as I see fit and try to keep my answers to about 500 words each. Remember, I failed my entrance exam, but that was about 15 years ago.

Anyway, these are what I now deem to be appropriate responses.

1. Outline the function of feasibility studies on projects.

Feasibility studies are of fundamental importance to a project's viability. It's okay to have a vision, but can you afford it? Once an architect has produced a viable drawing, or even an outline specification, it is possible to begin to evaluate both the capital expenditure required to construct the building in question, and also to give an idea of the life-cycle costings involved to both run and maintain that particular building. The feasibility study will outline how viable the project is, and also outline any particular problems or obstacles that may occur or that may need to be overcome.

The study may identify specific costs for items of work, such as the substructures (i.e. the below ground works, such as services and foundations), the superstructure (walls, roof, internal fit-out, etc), and such salient points as the IT and equipment requirements, including the bespoke electrical and mechanical features.

Furthermore, the feasibility study may comment, subject to the author's own expertise, on how good an idea the scheme actually is. They may, for example, say that the idea to build a residential apartment block with commercial units on the ground floor is especially suited to the locale, due to demand for such units in the area. It may even go a long way towards persuading potential investors to put their hard-earned cash into this particular development.

In all, a feasibility study should and could be undertaken for every development, from the large to the small, because it serves as a salutary pause for breath along the journey to realising any endeavour.

But that's just me!

2. How do you value variations against a schedule of works?

The schedule of work is the list of work items to be carried out. These will be costed prior to commencement on site. Any variation, therefore, is a separate item and needs to be priced based on its own merits. However, this must be done with one eye on any similar work items in the original scope of work. For example, if you are doing brickwork at £100 per metre squared, and the variation calls for an extra ten metres squared of the same thing, then this will be costed at the same £100 per metre squared price. Usually, that's the case, but not always. If that rate was based on a large amount of work being done in one continuous site visit, and the variation asks for a small amount to be done at a later date, then the variation needs to include for a return visit, possible mobilisation costs, any increase due to inflation, plus an uplift due to the small quantity involved in place of the economies of scale that were included in the original sum.

A variation that bears no relation to a previous work item needs to be costed on its own merits. Break it down into material costs, labour costs, a sum for any plant (mechanical, scaffold, etc), any supervision, plus a margin for overheads and profit. Get costs agreed with the client prior to commencement (*of major and fundamental importance!*). You can also negotiate. If the client thinks your price is too high, you may be able to shave a bit off. Or, you just have to explain to them the logic of your argument. I've allowed two men for three days, etc for the labour element. Your material costs came from your supplier, so what is there to argue with? You can't access the work area without a scaffold. So, again, what's their problem? And if you can't cover my overheads and allow me to make a little profit, why should I bother? Get someone else to do it.

If they don't have the money to cover the variation, and you're already quids in on the main contract works, of course try to help them out. But, generally, a variation should be priced as an individual, stand-alone item, harking back to previous rates where possible.

That's my answer.

3. What do you do if the lowest tender is also the lowest quality in terms of specification?

It isn't unheard of to get a phone call from a main contractor performing due diligence before placing an order. They like your price, but the job is due to start in two weeks' time. If they place the order with yourselves, do you have the labour available to commence at that time? You usually answer yes, even if you're unsure. You can move things (and people) around to accommodate the request. Especially if it means securing a juicy order.

If you're the people placing the order, you've usually (hopefully) received two or three or maybe four compliant tenders. Now you're faced with the enviable choice of which one to go with.

There are things such as *negotiated tenders*, where you absolutely know which company you want to carry out the work. You've used them before. They are the best at what they do. Why would you even consider opening the door to anyone else? In that instance, you come to some agreement about the price and you shake hands on the deal. One way to do this is to just say, if you trust that they will still be suitably incentivised to perform, we'll do it on a cost-plus basis. This means, at the end of the job (or at regular, say monthly, intervals if it's a long job) they give you an invoice for the material, labour, plant and supervision costs involved, and then you pay this amount plus an agreed mark-up of, say, ten, or 15, or 20%.

What to do if the lowest tender is also the lowest in terms of specification, i.e. quality? Well, you can outright reject it, or you can go back to the vendor and say, your price is fine, but I don't want a cheap and nasty job doing, so look at your figures again and give me your price to do your very best work using the very best materials. Will they still be the cheapest then? Who knows.

Essentially, you must check that every returned tender is on a par in terms of quality. Only then can you perform a proper analysis of which is the best in terms of value. It is no use saying, one subbie has priced for Levi's 501's, the other has priced on Asda's own brand. They aren't the same thing. Of course one is cheaper than the other.

So, you need to be clear as to what you want, give as much detail about the specification as possible, and then check the tender returns to see if they've all priced the same thing. If they have, and they are all equal as subcontractors in terms of quality of workmanship and availability to do the work, then by all means choose the cheapest. Just make sure you don't get a pair of Asda's own brand jeans if what you really wanted was Levi's 501's.

No wonder I never got in!

THE SMALL CONTRACTOR

By far the biggest employers within the construction industry are the small contractors. They are not merely the backbone of the industry, they are its flesh and blood. The next time you're walking down the street, look at how many vans you see advertising themselves as painters and decorators, plumbers, electricians, roofers, shop-fitters, and the like. They are out there in their hundreds of thousands.

I would classify a small building firm as a company employing between one and ten people, doing jobs ranging in value from five hundred to fifty thousand pounds. If you find yourself employed as a QS for a small contractor, you will be a big cog in a very small wheel. You'll multi-task, and you will find yourself taking a hand in every stage of a project. Personally, I find this stuff exhilarating.

Firstly, you'll get to know all of the trades-people who are doing the actual work. You'll organise the labour, sending so-and-so to whichever job on a daily basis. They'll phone you up with problems all day long. They need materials. You'll have to organise those too, so you'll form relationships with suppliers and builders merchants.

You'll deal with the clients face to face. You'll be their point of contact. You'll basically be the focal point for everything that is happening. If you like doing a little bit of everything, including project management, then this is the place for you.

Most small contractors have a body of work that is comprised of ad-hoc private work: people picking your name from the phone book or internet, or maybe spotting one of your vans, or by word-of-mouth recommendation (if you're doing a decent job). These small companies will also usually have a permanent contract with an established client such a local authority or housing association or a larger contractor. This will be their bread and butter. They'll have a steady workforce to feed, so they don't want to rely too heavily on random enquiries from the public.

In my experience, these standing contracts have often been won through a bit of schmoozing. Imagine, the surveyor or contracts manager for the client (e.g. the local authority or the large contractor) is handing out millions of pounds worth of work a year. These small companies are vying for his or her attention, and the surveyor or contracts manager can sometimes be charmed or otherwise persuaded to add your small company to their tender list. I'm not saying that this is a universal thing, it's just my opinion and – in my experience – it's something that goes on quite a bit. Don't be corrupt. Just don't be surprised if you find your employer wining and dining your client at the races or at a football match. It happens. Don't be corrupt. That is of fundamental importance, but don't be naive either.

A problem in dealing with bigger contractors, when you are a sub-contractor feeding on the back of the larger contractor's workload, is that you will probably experience the bully factor. That is, they'll expect you to be grateful to them for throwing you a bit of work, and you'll be expected to dance to their tune and put up with late payments and the like. Don't stand for it. Use all of your charm, communication skills, and powers of persuasion to get them to play fair. They sometimes work on the old adage that 'might is right.' It's not. 'Right is might.' If you're in the right, then stick to your guns, and get them to see sense. They won't want to lose you if you're a good sub-contractor, and if you allow yourself or your firm to be bullied, you'll end up as a victim; and no-one deserves that. Remember, 'right is might'!

I once joined a small contractor working on a prestigious contract for a Facilities Maintenance company. We were a subbie to the FM company. We had to refurbish a small office unit, and our contract value was about forty thousand pounds. With work well under way, I spoke to the FM company's QS and said something along the lines of 'I won't bother with an interim valuation, I'll just apply for the full sum of forty grand on completion.' His response surprised me. 'We don't work like that,' he said, meaning that I wasn't going to get the full forty grand. He was going to pay me whatever sum he wanted to. He was the client, the big boy, the bully, and I would just have to accept whatever value he put on the scheme, regardless of our contract and the actual contract sum.

I was shocked (I shouldn't have been. I was probably just naive). I thought about what he had said. We had signed up to do certain building works for a certain sum. This was the contract, enshrined in English law. How could he possibly ride roughshod over English law? Were he and his company somehow exempt? If so, what other laws was he exempt from? Could he commit murder? I wished I could at that point. The answer was that he was just trying to see if he could bully me into accepting whatever figure he decided to give me. I didn't waiver (and nor should you). The law was on my side. Right is might, remember.

When I was studying law at Uni, as part of my quantity surveying course, I always used the same approach. The tutor would set out the legal argument. So-and-so did this to someone. Someone complained. So-and-so denies it. How did the courts decide?

I would listen to the argument and think, 'Where does 'right' lie in this argument? Who is in the right, and who is in the wrong.' Once you've chosen the side you think is ethically in the right, all you have to do then is find similar case precedents to support your position, present them to the judge, jury, arbitrator or whoever, and hopefully you'll win the argument. I think Gandhi, who had previously been a lawyer, once said 'If you're in the right, the law, nine times out of ten, will come to your aid.'

Obviously, in Ghandi's example, that still leaves a one in ten chance that, even though you're in the right, you may not win your case. I would put that one in ten loss down to a failure in communication. The other side presented better. Even though you had a reasonable case, you failed to present it with sufficient clarity to sway the person making the judgement. Be clear, be concise, and communicate effectively. I've said it before, and I'll say it again, the role of the modern Quantity Surveyor is fifty percent communication. Know your facts, keep your nose clean, and communicate effectively. You never know, you may just get ten out of ten, in which case your services will be valued and even more in demand.

THE MEDIUM-SIZED CONTRACTOR

The medium-sized contractor is one employing between ten and fifty people, working on contracts from around ten thousand to a million pounds in value. As a QS for a contractor of this size, you'll find yourself occupying the middle ground. You'll be working as the main contractor for some schemes, and probably procuring subcontractors yourself; and then you'll also act as a sub-contractor yourself on larger schemes working for even larger contractors. You'll probably be both a boss and a subbie in equal measure and, as you'll usually have a handful of schemes running at any one time, you'll be expected to float effortlessly from one situation to another.

The medium-sized contractor will probably have their own buying department; therefore you won't have to do the procurement of materials yourself. You'll do the take-off (i.e. the measurement of the materials needed) and then pass that information on to your buyer, who'll place the actual orders.

The project that you're working on will have its own Contracts Manager or Project Manager, and you will be their right-hand man or woman. You'll run the scheme together, effectively being the management team for the scheme. You'll be concerned with the financial aspect, and they will be concerned with its practical aspect, i.e. the programming and the labour element. As one Contract Manager used to say to me, 'You do your stuff, I do my stuff, and then some of the stuff we do together.'

For example, a lot of the Contract Manager's working day was taken up by being asked to consider variations to the scheme. Once you're on site and performing well, the client is likely to ask if you can do a few other jobs while you're there. 'Here's another building that needs refurbishing.' That sort of thing. The bad QS sits in the office, and doesn't know what's going on out on site. The good QS responds to the client's request, looks at the job that needs pricing, gets a price back to them, and wins the additional work, adding value to the contract. The good QS will actually pay his own wages. You're a benefit to your employer, so make sure that you're remunerated accordingly.

We're all busy people, and none of us are geniuses, and precious few of us are paid what we are worth, therefore don't be surprised by the fact that things go wrong, or they don't get done right, or that projects fail, because it happens all the time. Very few jobs go from inception to completion without getting stuck in the mud at some stage. When a job goes 'live' on site – when it stops being a drawing and a design meeting consultation – and it's now happening in the real world, there's usually lots of gaps that appear in the project.

For example, you know you need a bricklayer, you know you need a joiner, you know you need a roofer. Those packages of work have been placed. But what happens where those bits of work meet up? The brickie has finished the walls, but before the roof trusses go on, we need some joinery doing to level up the wall-plate. Who's doing it? It's not in the brickie's package. It's not in the roofer's package. It's not in the joiner's package. The roof trusses are all coming pre-assembled, so it's not in their package either. These gaps all need to be plugged, and there are usually lots of them. If you're a QS for a builder or a subbie, grab as many of these additional packages as you can, and you will be swelling the value of your works order. Sometimes, you have to say I can't. We can't. We don't do whatever it is that they're asking for. But if you can, you'll become their go-to contractor of choice, because you like to say yes, and you're helping them fill the gaps and complete the scheme.

There's another area where the QS can add great value to the project, and to their Contract Manager, and that is procurement. While you may have a buyer or buying department available to you, it is important to get your orders placed early or at the appropriate times. I have seen so many projects fail because orders for materials or subcontractors weren't placed on time. *Don't leave it to the last minute. It is an absolute recipe for disaster.*

Say, for example, your project is running brilliantly. You're doing an internal refurb. You're putting up a thousand square metres of new partition walls. You can see the finishing line for the scheme. You're sailing to success. But you haven't yet placed your order for a plastering subcontractor. You need them on site next week.

You pick up the phone and speak to a couple of local plastering firms. You've used them before, or they've come recommended, or they're on your own company's list of preferred suppliers. Guess what, the firms you speak to are all keen to get the work, but they're mad-busy at the moment. They can't possibly start for two or three weeks.

All of a sudden, you've got a two or three-week delay on your project, because until the plastering has been done, you can't do the decorating, you can't do the second-fix plumbing and electrics, or the tiling, or install the fixtures and fittings, or do the ceilings.

Be ever watchful of the coming events and place your orders early for subcontractors and materials. If you do have a buyer or buying department, give them the list of what you need in good-enough time to allow them to do their jobs properly. They won't appreciate you dropping stuff on them at the last minute and then making a fuss because you needed it yesterday. They'll think you're difficult, or an idiot, and they'll be right! And because you're a difficult customer, they'll probably put you at the back of the queue anyway. You won't get the service you need.

The good QS in the office, who doesn't rant and rave at them, who doesn't make their lives difficult by putting them under pressure with late requests for goods and services, they will be at the front of the queue in terms of the buyers' response times. Conversely, as a bad QS, you'll be on the pay-no-mind list, and your project will be compromised accordingly. And it will be your fault. So procure well and procure early.

Another area where you may make a contribution is in the choice of contract. There are several to choose from at the start of any scheme. These are standardised contracts provided by the JCT – the Joint Contracts Tribunal – and which one you use is usually determined by the size and nature of the project, i.e. a small, intermediate, or main contract, or a design and build contract. For Civils schemes – usually large infrastructure schemes such as roads or bridges – you'll most likely use the NEC form of contract.

The terms and conditions, the clauses included and excluded, and the size of any penalties to be included, are all determined between the client and contractor, usually in a face-to-face meeting (or several). Here again, your communication skills will be vital. Don't sign up to anything that you're not happy with. This is the moment to ask all your questions and to seek clarity on any points that you're not clear about.

If you sign up to complete the project at the end of June (say) with liquidated damages at ten thousand pounds a week thereafter, don't go crying to the client or an adjudicator when the project runs over and you've got to cough up the cash or see your final account sum diminish to reflect the damages. You signed it, you're stuck with it; so think about all the implications of each and every clause you sign up to.

If you can live with it, sign it. If you can't, don't.

And don't be afraid to negotiate on these individual points. Sign up to two grand a week, or whatever, in liquidated damages. Just don't sign up to anything that you're not happy with. An absolute recipe for disaster is to have two sides with different expectations or interpretations of what is about to unfold. Don't be afraid to speak clearly. It's not dumbing down. It's just common sense.

I used to work in an Estimating department for a national roofing company. Each tender we would return would be slightly different, owing to the nature of that particular scheme or client. I used to tailor each of my clauses individually. If you're putting roof vents into your new roofs, at sixty quid a pop, at two per dwelling, for a scheme of a hundred houses, that's twelve grand's worth of roof vents. If you've excluded them from your quote, write 'This quotation does not include for any roof vents. Should you require roof vents, they are sixty pounds each'.

What could be clearer than that?

If you get on site and the client, a large house builder, says 'I want two vents on each roof, you can say 'Great. They're sixty quid each'. The contractor will probably say, 'I expect these as standard. You better put two up on each roof or I'll never work with you again'. You stick to your guns. What part of 'This quotation does not include for any roof vents' did they fail to understand? You're in the right, but to avoid any confusion, if it looks like your tender has been successful and you're in negotiations to start on site, point it out to the contractor. 'You know our quote doesn't include for any roof vents.' They might say 'Oh, I didn't realise.' They'll have to add twelve grand on to your quote.

Are you still the lowest? Maybe not. You might lose the contract. Better that than to start the job and end up in court. That will cost you time and money and damage your reputation. Better to be clear and unambiguous. Just speak in plain English. Your client will like how you roll. They'll see that you're straightforward. They'll like doing business with you because they'll know where they stand.

Similarly, when dealing with subcontractors of your own, be clear about what you want and what you expect. I want so-and-so done. How much is that going to cost?' The subbie gives you a price. You accept it (or negotiate with them and then accept). The job gets done. *As long as you're happy with it, you pay them the agreed amount. No animosity, no confusion, and no day in court.*

Guess what, the next time you pick up the phone to them, they'll come running. No waiting around. They'll do their best to accommodate you. We're back in the world of procurement. There was no delay to your scheme because your subbie came running when you needed them. It was all about relationships, all about communication. Information given in a timely, accurate, and clear manner. It's not that hard, is it? Believe me, the bad QS finds this extremely difficult to grasp, and doesn't understand why all of their schemes are running late and not performing properly. The good QS does good communication all day long. It's part of the package.

I'll tell you about another job I worked on. This is a prime example of how NOT to run a building project.

The scheme was to construct 68 new-build luxury apartments in a tower block in Manchester. On my first day on the job, I was greeted by the site agent in his office. I asked him when the project was due to be complete. Having passed the building on my way into the office, I could see that there was still a lot of work left to do, probably six-to-twelve months' worth.

The site agent told me it should have been completed six months earlier. He then explained that he had spent the previous evening up a ladder fixing a leak from the roof into the building. When he described the nature of the leak, I said it sounded more like a design fault than defective workmanship, and how come he – the site agent – had been tasked with fixing it? 'If I don't, *he* shouts at me!' he said.

Oops!

Who's '*he*'? I wondered.

All of a sudden, the Project Manager came barging into the office and started berating the site agent about a set of keys that he needed, and where they were. It was no way to speak to any member of staff, and certainly not your right-hand man. I could immediately see where the problems lay with the scheme. They lay with the man at the top, the so-called project director, the captain of the ship.

I introduced myself and was then given my own desk and office. I started to read through the emails in my inbox. There was a glazing contractor who was threatening to take our firm to court. He had tried everything to resolve the situation amicably, and if he didn't hear back from us within the week, solicitors would be called.

I found the file containing all of the correspondence (it's always great reading your boss's emails!), and I began to acquaint myself with the background to the dispute. Essentially, the glazing contractor had fitted the windows and now wanted paying. That all seems pretty straightforward, doesn't it? Well, not with this guy in charge.

He had used every trick in the book in order to avoid paying. The correspondence trail was fascinating, and damning. He was citing non-performance as a reason to deduct damages and reduce the bill (to nothing!). He would fire off letters saying 'your men left the workplace untidy, so I'm going to charge you.' The glazing contractor would apologise and say they had spoken to their operatives and would ensure they cleaned up properly after themselves. When they proceeded to tidy the site, the Project Manager would then fire off another letter saying they were filling the skips up too quickly therefore he was going to have to charge them for the additional skips! The glazing contractor couldn't win.

Another letter would come in, asking for their due payment. The Project Manager would say he wasn't going to pay them because the glazing contractor hadn't placed enough men on the job, or that they had left the site for a couple of days, which had a negative effect on the programme. The glazing contractor pointed out, quite reasonably in my view, that they had sent sufficient men to site and had installed windows in all of the available windows, but that the brickwork had been built to incorrect sizes on certain floors of the tower block; a fact which they had conveyed to him immediately, and they had only left the site to give the bricklayers time to put their work right. The glazing contractor couldn't be expected to have their men sitting there doing nothing for a week. They had returned to the site as soon as they were informed that the window openings had been re-sized and that they could now install the windows.

The next day, when I was at my desk, one of the partners of the construction company came to visit. I introduced myself and told him of my concerns that his company was about to be dragged into court. I gave him my advice that, having read the correspondence file, any arbitrator or decision-maker was bound to find in the glazing company's favour. At every turn, I informed him, they could be said to have acted reasonably, and at every turn, our own company could be said to have acted belligerently, in my opinion. He took it in his stride (he did, in all fairness, seem like a reasonable and capable man) and told me that the matter would be settled out of court. That was the end of that. I wasn't there long enough to find out. I left after a week!

Before I left, the Project Manager then told me that the bricklaying contractor had walked off site, never to return. They wanted no more of the project, or of him. He asked me to get him another bricklaying subbie. My immediate thought was 'I'm not going to bring in anyone that I know!' Firstly, there was a real danger with this man in charge that they wouldn't get paid. Secondly, they'd have to deal with his nonsense.

I'd only taken the job on a temp basis while I awaited the outcome of an interview I'd had for a job that I craved with a QS/Claims Consultant. It had the element of contract and construction law that I wanted. Anyway, I was told that I'd got the job and could start straight away, so I only had one week with the company building the tower block with the toxic Project Manager. Apparently, it was a good thing that I left because he was going to sack me! I think my attitude failed to impress him, refusing to jump when he asked and not paying him the respect that he thought he was due. I'll stick to my guns on that score.

A couple of major points to make here about this scheme.

One day (during the week that I was there), I walked onto the site to see the work in progress for myself. You have to get as close to the work as possible. The more you have the project in your DNA, the better able you are to contribute and communicate, to problem-solve, and to talk about any financial aspects with your boss, the client, or whoever. Get it into your DNA.

Anyway, on this site walk-around, I was amazed by the lack of operatives on site. They were six months behind on a major development to construct 68 luxury apartments in a high-rise city-centre block. Except the place was like a ghost town. There was hardly anyone on site. Eventually, I found about four guys actually carrying out the work. I introduced myself as the new QS.

'Have you met Mister Personality yet?' one of them asked me.

A job that was six months behind should have been inundated with labour in order to get it finished. Unfortunately for the Project Manager, this was at the height of the building boom, in 2006, and no one wanted to deal with him. Why work on his scheme when there was plenty of other work around? What would you do? Work, and possibly not get paid, for a cantankerous employer, or go elsewhere and work with a reasonable employer and pick up your wages every week? It's a no-brainer. So the site was practically empty, and the project was six months in delay, with a further six or twelve months to go to completion.

Finally, the other devastating outcome was that, if the project had been completed six months earlier, as it was meant to(!) the economy and housing market was booming and the developer would have had no problem selling those apartments. But, by the time they were finally finished, the bottom had fallen out of the market and it was virtually impossible to sell them, meaning almost certain insolvency for the company. That's what a bad Project Manager can do. They can kill your company. Remember, as a QS, one of your five main skills is to be a Project Manager. Be a good one. The bad ones kill the company.

THE LARGE CONTRACTOR

The large contractor, i.e. companies employing more than fifty people directly, working on contracts of a million pounds and more in value, are a relatively rare beast in the construction industry. You can probably name a dozen or so yourself, the Kiers, Balfour Beattys, and Laing O'Rourkes of this world, as well as the major house-builders. These are the companies that can take on large projects such as airports, hospitals, and highways. Even then, it is not uncommon for these companies to 'partner-up' to meet the capacity required of a large construction scheme.

Working for a large contractor means that you will be a small cog in a large wheel. You'll probably get to work on some prestigious projects. These are the sort of thing where you'll probably drive your friends and family wild for the next few years when every time you drive past the place, you'll say 'I built that.'

What you won't get to do, in all likelihood, is lots of different things. You won't be multi-tasking with procurement and project management. You'll be a Quantity Surveyor, pure and simple. You might be looking after several sub-contract packages, but you'll be pretty much doing the same thing with each: placing the order, monitoring the work, and valuing their accounts on a monthly basis. This can be interesting and exciting enough. After all, the project will probably be a huge and possibly glamorous one. It will be interesting watching all the separate components of the project coming together. It will be important to keep a vision of the whole project in your head. You'll be responsible for placing large orders, and you'll be dealing with large sums of money. That's a great deal of responsibility.

I've worked with a couple of large contractors on some very large schemes. One was a three hundred million pound hospital scheme in Scotland. Another was a one hundred and fifty million pound project for the Ministry of Defence. I've also been a Property Manager for five hundred million pounds of programmed maintenance for a large commercial property portfolio. I found it all interesting, and challenging, and rewarding in its own way, and these projects will be discussed elsewhere in the book.

When I have sub-contracted, say for a medium-sized contractor, I found it quite interesting to deal with the main contractor's QS. You have to stand your ground. You'll usually find that they are quite entrenched. The large contractor usually has a company ethos that is embedded throughout the workforce. You'll be expected, to some degree, to dance to their tune. It's ultimately their project. You're just a small part of it, and you'd better perform or you're going to be jettisoned from the job.

I recently worked as a site-based QS on a four million pound sub-contract as part of a huge scheme for the MOD. The main contractor's QS was telling me about the rules of the game, as he saw it. Having previously been instructed by my own boss not to be bullied by anyone, I responded honestly to the law being laid down in my direction. Basically, our sub-contract package was growing by the day. As we were performing well, the main contractor was throwing more and more work in our direction. Their QS wanted to make sure that we were going to jump when he wanted and be grateful for his benevolence.

My response was, 'You tell me what you want, I'll give you a price for it, and if you like the price, then you give us the go ahead. It's as simple as that.' His response was, 'Oh, no. It's not that easy.' But why not? It can be as simple as that. It *should* be as simple as that. The only reason not to make it as simple as that is if one party is trying to procure some advantage by muddying the water and sowing confusion.

To my mind, there is no benefit to either side to work in that manner. There might be some short-term advantage gained by one party, but when the penny drops for the other side that they've been cheated, all the goodwill goes out the window and ultimately the project suffers in the end. The client will end up paying somehow. The next contract variation will be loaded. The next favour requested will be refused.

I had a subbie of my own on the same project. They gave me a quote for some work, and I pointed out that they had missed a vital component of the work and asked for it to be included. My own employer asked why I was asking the subcontractor to increase their quote. I answered that if they went ahead and installed only the elements they had quoted for, we wouldn't be able to retrospectively fit the missing parts. This would have a detrimental effect on the programme. I might have been accused of deliberately misleading them because I could clearly see that something was missing from their quote. We'd have ended up squabbling in front of the main contractor. None of this would have been good.

It would have damaged our reputation. It was much better to nip it in the bud early. That way, the job would be getting done properly from the beginning, which is never a bad thing. Sometimes, trying to save a couple of grand can be counter-productive. Remember, the right way is always the right way. So, whether you find yourself working for a small, medium, or large contractor, you'll find the roles slightly different in each environment, but a good QS is always a good QS, and that's what you should aim to be.

PRIVATE PRACTICE SURVEYING

Private Practice surveying, where you will be a Private Quantity Surveyor, or PQS, is also often referred to as client-side surveying. In any contract, there will usually be two sides to the arrangement. Think of it as a handshake. Two people shaking hands on a deal. On one side, you have the client. On the other, you have the builder or contractor. Say, for example, a local authority wants to build a new school. They want someone to build that school for them, because although the local authority will be the end-user, they specialise in teaching children, not in putting up buildings. So we have two sides to the deal: the client, and the contractor.

The contractor will have a contracts manager or project manager overseeing the scheme, and a Quantity Surveyor overseeing the finances of the whole operation, not to mention a team of specialised trades who will carry out the physical work. Similarly, the client will also want to know that they are getting value for money, and that the work is being carried out to the right specifications and at the right price. So, the client will also employ a QS. The Quantity Surveyor on the client-side will be the PQS.

The contractor's QS will submit an application for payment on a weekly, fortnightly, or (more usually) monthly basis, and the PQS, on behalf of the client, will check the application before payment is issued.

For example, the contractor's QS might apply for one hundred percent of the brickwork element of the contract. The PQS will visit the site and maybe notice that only ninety percent of the brickwork is complete. He will reduce the payment accordingly and pay the correct amount, protecting the client's interests.

Quite often, and to save any misunderstandings, the PQS and the contractor's QS will visit the site and walk around together. They can then value the work at the same time and agree on a percentage there and then. They both look at the brickwork. The contractor's QS says 'I think that brickwork is complete so I'm going to apply for one hundred percent in this month's application.' The PQS says 'Don't be daft. There's a whole section there that is incomplete. I think the brickwork is only ninety percent complete'.

Maybe there is work in progress. Maybe, by the end of that week, the brickwork is going to be 95 percent done. You both agree to value the work at 95 percent. After all, you have to be able to negotiate, and to compromise, and to value good relationships for the benefit of the project.

Don't try to overkill a good opponent. You want these people to still be on site next week. Don't lose sight of the big picture. If they're doing a good job, if you can see they're making progress, and if you know that they're going to reach that percentage figure in a few days anyway, go with the flow.

The contractor may have only applied for fifty percent of the plastering. You're more than happy with that. Last week, he did a little variation for you. It was a favour. If you have good relationships with these people, if they're playing ball with you, *then don't take the ball home with you.* Keep them on side by being flexible and reasonable. If you know that the plastering and the brickwork combined, in your estimation, comes to a figure closely matching their application for payment, then pay it. It will do you a lot more good than counting every penny and arguing over every item. We're all busy people. We're trying to finish this building for the client. Don't slow things down by taking everything to three decimal places. We're not NASA. We're QS's. Good relations, a team-ethic, and a reasonable attitude will take you and the project a lot further than fighting over every penny.

That said, you can't overpay. Don't be profligate. One, it's not your money; you're there to protect the client's money, so remember that. Two, if you pay a contractor for a hundred percent of the brickwork when they're only ninety percent of the way there, then what happens if they go bust, which sometimes happens. You have to get another contractor in to finish the brickwork, and you've already spent the money to do that element of the work, which is irresponsible.

So, keep a cool head, value the project reasonably, and do your best to keep everyone on side for the benefit of the project.

Don't sweat the small stuff, always keep an eye on the big picture, and protect the client's interests by monitoring costs and performance, but don't let your own contribution be a bar to progress. Ultimately, you will have to settle the final account with the contractor. Hopefully, this too can be done amicably and sensibly. Personally speaking, I believe in fairness to all. If the contractor has delivered exactly what they signed up to do, then you have no reason not to pay them the contract sum.

Remember, if done right, a contract is nothing more than a handshake. The only difference in your role as a PQS is that, client-side, you're on one side of the handshake, and as the contractor's QS you'll be on the other side of the handshake. But it's the same handshake, the same deal, and you should both want the same thing: the building completed on time and on budget. If that's not what you want, if that's not what you're there to achieve, then you shouldn't be involved in this particular handshake.

I worked in private practice for three and a half years. I was an assistant QS at the time, and I worked with some great QS's. One difference between being a contractor's QS and a PQS is that, working for a contractor, you're in amongst the actual build-team, the builders themselves. It's a lot more rough and tumble, the language quite raw and direct. It's real-world stuff. As a PQS, you're talking to the client. Maybe they're a millionaire property developer or a local authority. You have to mind your language, although there's nothing wrong with showing a little bit of personality as you go about your business.

As a PQS, you'll probably work in a nice office environment. You'll hardly hear a swear word or a word spoken in anger. You certainly will as a contractor's QS, dealing with the nitty-gritty stuff that goes on out on site every day. What sort of a character are you? Which do you think you'd prefer? Can you hold your own in an argument? If you can handle the rough and tumble, you'll be fine as a contractor's QS. If you prefer a more genteel existence, maybe you'll be more suited to the life of a PQS.

Another thing I found, as a PQS, was how fastidious it was. You'll refer to measurement books such as SMM7 (as it was in my day) or the new National Rules for Measurement (the NRM).

Everything is measured to the nth degree. I used to hate it. I used to refer to it as disappearing down the plughole.

You measure the internal walls of a building. Fair enough, you need to tell the builder how many square metres of stud or block wall to allow for, in their tender. Then, the measurement book will ask you 'How much of this walling is between three metres and three point three metres high? How much is between three point three and three point six metres high', and so on, and so on. I'm thinking, 'I've already given you the overall amount. Why ask me to break it down further?' It just wasn't for me.

It's fair to say, if it was my money, if I were the client, I'd probably welcome the exacting attention of a PQS service. I'd want every penny accounted for, to two decimal places. As the guy doing the actual measuring though, I thought my talents were better suited to considering the big picture, not looking down the plughole. I like broad brush-strokes. However, if you prefer structure, sometimes infinitesimal structure, then maybe you're better suited to Private Practice. Personally, I've done it all, and I like it all, but I know that I'm better suited to some parts than others. You will be too.

PROJECT MANAGEMENT

The project manager, or contracts manager, will be at the forefront of any live project. They'll do what their title suggests, managing the project or contract.

We've discussed what a bad manager can do to your project. A good project manager can lead you to success. And the good news is, as the Quantity Surveyor for a particular project, you'll be the project manager's right-hand man or woman. You'll be their consigliere, their counsel. You'll probably share an office together, and you'll be right at the heart of the decision-making process.

All the important choices will be made right there in front of you, probably with some input from yourself. There'll be no secrets from you. It will all unfold before you. And that's exciting. This is your opportunity to make a powerful contribution. You should form a tight-knit partnership, like Starsky and Hutch or the Blues Brothers. You should talk to each other at every opportunity and overcome problems together.

Your role will be to advise on all financial matters concerning the project, but there will be a great deal of cross-pollination between your individual responsibilities. You'll be a team. *Everything* that happens within the scope of the project affects its financial performance, and vice-versa. Every financial decision made regarding the project affects its performance. Therefore, you have to work together to achieve the best results.

You might be falling behind in terms of programme. The project manager might suggest doubling-up on the labour or working weekends, or bringing in an outside contractor at a premium to speed things up. Have you got the money for this acceleration within the budget? You have to weigh that up against the cost of any penalties for delivering the project after the completion date. It's a balancing act, and if you don't work closely together, then you're probably both going to lose out. The programme will be affected and your finances will be affected. Work as a team.

The Contract Manager or Project Manager will have a different skill-set to you, and they will have different responsibilities regarding the Health and Safety aspect of the work being carried out on site. They will have had to have undergone the requisite training to acquire the qualifications to manage a project on site.

As the QS, the rules are less rigid on this, though it may be something that you wish to undertake. Essentially, as a QS, and as discussed previously, you'll be called to either work closely with the project manager or, if you're working for a small contractor, do a bit of project management yourself. Again, your construction knowledge and your communication skills will come to the fore. Can you motivate the workforce to give a little extra? A good PM can. Do people want to work for you or, as we saw on the Manchester tower block scheme, are people running for the hills to avoid you?

The five different tasks a QS performs are chiefly

1. Estimating
2. Procurement
3. Valuations
4. Negotiating
5. Project Management

These will be discussed in greater detail later in the book. For now, it's safe to say that the Project Manager or Contracts Manager is fundamental to the project's success, and they need a good QS at their side every step of the way. You'll be a sounding-board, a soul-mate, and an advisor on all matters financial. It's up to the two of you to make the project run smoothly and to deliver the right financial result at the end.

TO STEAL OR NOT TO STEAL!

Stories are rife that the construction industry is crooked, and the Quantity Surveyor, as holder of the purse-strings, is prone to more suspicion and innuendo than almost any other professional in the industry. You can understand why. The QS has access to all of the costs and is responsible for the distribution of all of the funds contained within the budget. It's their job to dish out the cash.

Any book on Quantity Surveying will have a section on ethics, and rightly so. As this is a subjective account, I can only speak from my own experience and give my own advice. The total amount of all the money that I have taken in bribes or coercion in 20 years of QS'ing is a big fat zero, although I did once allow myself to be treated to a complimentary weekend at a hotel. We had been staying there for over a year, myself and about 20 tradesmen, working on a large construction project many miles from home. Four nights a week we'd stay at this hotel, eat our meals there, have a drink after work, play a few games of pool or watch the football on the television.

Occasionally, if there was a problem in the bar or with a bill, I'd be called from my room to help sort out the situation. As I was the manager on site, this responsibility carried through to the hotel. For this additional hassle, the hotel offered me a comp'd weekend. Once the job was completed, and as I happened to be heading down to that part of the world on vacation, I took advantage of the offer. That's the only outside benefit I've ever accepted for doing my job, but you do occasionally hear of quite unbelievable amounts of money changing hands, though I have no proof of this. It may just be an industry myth… although I doubt it!

I was once told that a QS received fifty thousand pounds as a bribe to place an order worth over a million with a particular contractor. This may be apocryphal, but I'm sure it goes on to some extent. We're not talking FIFA-level corruption here, but there must be some temptation when people are awarding contracts involving significant sums of money.

I have always offered the same advice to young professionals entering the industry or studying for their QS qualifications. There will be opportunities for you to befriend a particular sub-contractor, to get a little pally with them, and maybe to give them the nod about where they should pitch their tender in order to win the contract. I should imagine that the subbie would be quite grateful for this information and would want to reward you. *Don't do it.* You can communicate and negotiate with people that you have a good relationship with. You can advise them, encourage them, and cajole them, but don't defer to them and *don't allow yourself to be bought.*

If you're a QS, you're going to be earning a pretty good salary your whole career. That's something that you can build a life around, raise a family with, use to attain a mortgage. You've got that salary every week or month for the rest of your working life.

Now, imagine if you're in the pocket of a particular sub-contractor. All of a sudden, the client, or your employer, starts to notice this cosy relationship. Your subbie is winning lots of work from you, usually coming in just a fraction below the tender sum of their nearest competitor. Your client or employer starts to look closely at the car you're driving, and the suit you're wearing. Their ears prick up when you say you're off to the Caribbean for two weeks with the whole family for your annual holiday, especially when your employer is off to Wales for the week! Something's not quite right. There's something going on. Even if there isn't, there'll be the suspicion that there is. The role of the QS is to hold the purse-strings of the construction project. Would you hand your purse to someone that you suspected of being corrupt? I doubt any of us would.

Warning: *suspicious activity can damage your whole career.* If you're taking home two thousand pounds a month, every month, for life, do you really want to jeopardise that to make a thousand pounds on the side?

I remember one time being sent to the site to meet a joinery contractor who had been installing windows on a large office refurbishment. I had valued his account at eleven thousand pounds. He had put in a claim for fifteen thousand pounds. We met on site to walk the job and to finalise a figure that we could both agree on. It turned out that he had made an error in his calculations. He had claimed for installing 242 metres of a particular item. When he checked his notes, his actual measure was 24.2. He had over-claimed by accident. It was a simple and honest mistake.

We chatted amicably, and we were getting quite pally. It crossed my mind, that if I so wished, I could have easily ignored the mistake and valued his account at, say, thirteen thousand. No one back at my office would have questioned it. We weren't talking millions of pounds so it wouldn't have been scrutinised in any great detail. The joinery contractor, knowing the true value of his claim to be eleven, would happily have accepted thirteen. Then, we could have arranged to meet in some out-of-the-way bar, or some subterranean car park, and he could have given me a thousand pounds in cash in a brown envelope. He'd have been left with twelve thousand, so he would have benefitted to the tune of a grand and so would I.

But what if someone from my work, or another contractor, or the client had happened to call into that pub that particular night? They'd have wondered what I was doing meeting a contractor in this way, and would probably have grown suspicious and questioned my integrity. And they'd have been right.

Above all, though I sensed opportunity, I also knew that I had signed up to be a QS for life, that I was a construction industry professional, and that I had responsibilities. It wasn't worth risking all that for a thousand pounds; so I had a laugh and a joke with the joinery bloke, and valued their account at eleven thousand pounds, and my own career at a lot more.

You may try it yourself and get away with it. All I can say is, when there is an economic downturn and it becomes like a game of musical chairs out there, with everyone trying to grab a job before the music stops, those QS's that have a whiff of suspicion about them will be the first ones to go in any cull. And, if you know yourself to be deserving of suspicion, you won't have any moral high ground on which to stake your claim to be allowed the opportunity to stay.

But let's not be naive. I once joined a small contractor and was given the task of running a project that had already been tendered for successfully, and which was about to start on site. I was about to procure the materials and the subcontractors needed to complete the programme. In order to familiarise myself with the scheme, I began to read through all the contract documents. In amongst this cache of information were the details of how the contract had been won. We were one of three companies bidding for the work. It was a small slating job to the roof of a school. The third and highest of the tenders came in at twenty-seven thousand pounds. This seemed a fair price for the work. The second tender came in at twenty thousand pounds precisely. Our tender came in at fifty pounds less!

Immediately, I sensed the situation. The highest tender was probably a fair price for the work, if not a little high. The second tender was probably submitted by a contractor who was growing a little exasperated by the whole deal. He'd probably been submitting tenders unsuccessfully to the same surveyor, and was beginning to smell a rat. He probably thought, the job is worth twenty-five grand, but I'm going to go in ridiculously low, either to win it, or to flush out the hand of the surveyor awarding the work. The second contractor – on hearing the results of the unsuccessful tender bid – probably declined to bid for any further work. He was being used as a pricing exercise. The surveyor, who was obviously close to the company I worked for, could show his own employer that he was fulfilling his duties diligently. He had gone out to three different companies to obtain prices, and he'd awarded the contract to the cheapest of the three. What could be more straightforward than that?

However, I imagined that if we were to look at the history of previously-awarded contracts, a pattern would emerge. This sort of thing would probably be happening quite a lot. An auditor would quite rightly deem the whole affair suspicious and would question the relationship between the surveyor and the contractor who was winning most of the work. In all likelihood, that cosy relationship was expressed in some kind of remuneration – like cash or that family holiday. It's a fine line to walk. I'm sure that many people have crossed that line quite successfully throughout their careers. It's your choice but, like I say, *you're putting your career on the line every time.*

My advice, therefore, remains as it has always been. Don't be corrupt. You're a construction industry professional. As such, you've taken a kind of moral Hippocratic Oath. You're holding the purse-strings of any construction project, so do all that you can to avoid suspicion. Be open instead of secretive, communicative instead of furtive. Emails create a good audit trail. Keep a diary if you have to, which will at least help you remember what meetings and appointments you held, even if these meetings aren't minuted. Above all, treasure your reputation. Play it straight. If you don't, it will sully your career, and it will mean that you are denied opportunities to spend your career in the exciting world of construction. Be a QS. Be a good QS. Don't be a bad, corrupt, or merely incompetent one. The industry is crying out for good QS's. Are you one? I hope so. The rewards are fantastic and, by that, I mean the honest ones.

FREELANCING. PART ONE

The good news is, once you're qualified as a QS, you're likely to be in demand. When the economy is booming (fingers crossed!) you'll feel like that rock star or film star we mentioned in the introduction, because the phone won't stop ringing. Employment agencies will be looking down the back of filing cabinets to find old CV's for any QS that they can find. They'll want you that bad.

When I did my Master's degree in Writing, I asked my employer if, in 18 months' time, when I came to write my dissertation, I could take a sabbatical. They were a large company, with 3,000 employees. Surely, with 18 months' notice, they could accommodate my request. The subject I had chosen for my dissertation was so complex that I realised a 15,000 word thesis wouldn't touch the sides. I wanted to keep writing. I was going to turn it into a book.

The message I got back was, 'sure, sure, but I probably won't be your manager at the time. Take it up with the next person.' Then, a month before I was due to go, I was told that I couldn't go! The company had no one to cover me. They had put nothing in place to take account of my absence. I couldn't go.

Well, sod them, I thought. I had no kids. I wasn't married. I wanted to write my book. And so I left. A job I loved. Colleagues I adored. A job that, in three years, I had never wanted not to be there. Of course, there were days when maybe I'd been out the night before and wished I didn't have to go to work that day, but there was never a day when I didn't like my job. But I left. And I wrote my book. Six months to the day after leaving my job, I wrote 'The End' in my book. And then I went back to work.

This was where I had the week with the bad PM in Manchester. This was where I got the job that I craved in Claims Consultancy. I had decided to go down the route of Freelancing. The Freelance QS is an acknowledged and major part of the QS canon. Indeed, it is almost the Holy Grail. It is the ability to go wherever, whenever, for as long or as short as you or the client wishes. It is to be the Red Adair of Quantity Surveying.

For those that don't know, feel free to look up Red Adair now. For those who do, suffice it to say he wasn't just a literal fire-fighter but a metaphorical one too, going where he was most needed, called upon in a crisis. He was the right man for the job: skilled, willing, and able.

And this is what I mean by being able to show a bit of personality. At my interview for the job, I was asked where I had worked previously. There was a six-month gap in my CV. Why had I left?

I told the man that I had left to write a book. What book? Israel / Palestine – the complete history (if you're interested!). We had a chat, at the end of which, he said 'I think your book is going to sell a million copies. Do you want a job in the meantime?' So, it's never a bad thing to put yourself in there.

Actually, the subject matter of the book played right into the hands of the role for which I was applying, which essentially involved solving disputes between contractors and subcontractors or clients. It was all about gathering facts and presenting a narrative. There was a lot of research involved. You had to keep your analysis clear and concise. And, if I could sort out Israel and Palestine, I certainly stood a chance with So-and-So Contractors versus Mr and Mrs Smith!

Anyway, I freelanced for four years. It was great. The money is phenomenal. The exposure to different experiences is phenomenal. Basically, clients pay a premium for your services because they are desperate. You are available at short notice. Your contract could be for anything from a day, week, month, or year. The only downside (and it's a big one) is that you'll be the first one jettisoned when money becomes tight. They might still need you but, by then, they can no longer afford you.

Imagine if you had your own client list. Imagine if two or three or four people used you as their Quantity Surveyor of choice. They don't have the handicap of an employee on the payroll. You have the freedom to go where and when you like and divide up your workload accordingly. Being a freelance QS is effectively being your own boss while working for other people.

I worked for the Claims Consultants for a year. I learned and earned like I never had before. I left when they tried to send me on an assignment for a year that meant being away from home for all that time. I didn't mind it for a month or so. You'd be away from Monday to Friday. I'd come home to a stack of mail regarding my book. I couldn't fit my life into a weekend. I had a serious relationship on the go.

Worst of all, the boss had told me that this new potential assignment was only for a month. I had met the client personally and had been told it was going to last for a year at least. I felt that I had been deceived, and there was no way that I wanted to be away for that amount of time anyway. So I resigned.

There had been some great assignments along the way. I'll discuss one or two of them in the case studies later on. Suffice it to say, it was an eye-opener. They were like a cab-office for quantity surveyors. We'd all be sitting there, and then some desperate construction company would say 'I need a QS', and our employer would say, right, who wants to spend a month in Stirling, Scotland, or in Drogheda Bay in Ireland, or in Sheffield. And off we would go. At a premium. Fire-fighting with our QS skill-set to the construction-industry needy.

I then did three months for a curtain-walling contractor, and then six months as project QS for a £6 million new-build health centre, only a couple of miles from my home. It was great. Then, when the project completed at Christmas-time, I thought, I can try to find another assignment, or I can take a few months off. Everything I earn will be swallowed up by the tax-man now anyway, as I'd been earning great money as a freelance QS. I'll take the rest of the financial year off. I disappeared off to the South of France for three months. It inspired two novels. Not a bad lark, this freelance QS'ing.

Then, the economy crashed. Then I didn't have a job. If no one has any money to spend, and you're not an employee, then you really are on your own. If you are going to go it alone, get clients, (get good ones!) and be good at what you do, and therefore in demand.

In a booming economy, being a freelance QS is amazing. But, you're on your own. Whatever happens, no matter how bad things get, no one is coming to your rescue. Do you have your own clients? Can you work for yourself? Can you survive an economic downturn? For most of us, the reality is (as it certainly was for me), you search, and search, and search, and search until you find a job that may just bear some relation to your role as a Quantity Surveyor. In my case, I found a job in an estimating department. And I learned. And I had a laugh. And I'll tell you later on about my most recent stint freelancing.

In a perfect world, there's not much better for a QS than being freelance. But the world ain't perfect. Of course, you already knew that.

ESTIMATING AND TENDERING. PART ONE

As a Quantity Surveyor, you will be closely acquainted with estimating and tendering. Let's look at both of these aspects in detail.

What is an estimate? Outside of a construction context, you'd probably say that an estimate is a rough guess or a close approximation. It is not an exact figure.

In a construction context, an estimate is an educated guess, based on your knowledge and experience. Basically, if you're preparing an estimate, you're trying to get to a figure that closely resemblances the *actual* costs of carrying out the work. However, because your estimate is made in advance of the work being carried out, it can only ever be your best guess. How close those two figures match up, determines how good you are as an estimator. Get too many wrong, or by too wide of a margin, and your days as an estimator are numbered (no pun intended).

So, how to prepare an estimate? Well, there are certain publications that can help you, such as Spons or Laxtons, however these should be used as a rough guide or 'check' only. The best guide that you can have is to get to know your own tradesmen, or your subcontractors, and to base your price on previously done work. These are the best indicators that you can have. Say you've been asked to give a quote to build an extension to a house. The first thing to do is to break it down into its component parts; that way, you're unlikely to overlook anything and you can consider each component in isolation.

Of course, it's worth mentioning here that you have to look at the location first and foremost. Are there any access restrictions or site conditions that may affect how the work will be carried out? If so, you need to make an allowance for that element. If not, then the pricing should be pretty straightforward.

Allow for building the foundations, then the walls, and any windows and doors. What is going on the roof? Is it a flat roof, or a pitched roof with either tile or slate? Allow for plumbing, heating, and electrics. Allow for floor and ceiling finishes, plastering to the walls, and decorations. Are the walls being rendered externally?

You need to establish the scope of works.

Usually, an architect will have drawn up plans for the scheme and will list the items of work and the specification detailing the materials to be used. The client might be a millionaire and may desire gold taps. You *need to know* all this. If you price for standard items and the client and architect have specified something more elaborate, then you're going to have a headache somewhere down the line. You may lose all of your profit on the project, so be sure to know exactly what it is that the client wants so that you can price it accordingly.

The next thing to do is measure each item. So, what is the length of your strip footings? What is the square metre-age of your brick and block walls? What is the area to be plastered? The same for the roof. How many windows are there? How many doors, and what type? Maybe they're patio doors. What type of flooring? If the client wants a marble stone floor, this is going to be more expensive than a bog-standard vinyl floor. The materials are more expensive, and they take a lot longer to lay, so make sure the costs contained in your estimate reflect these stone-cold facts.

Once you have some measurements against each individual item, you can start to put in some rates against each one. You might allow £100 per linear metre for your strip footings. You might allow £100 per metre squared for your brick and block cavity walls. You might allow £20 per square metre for your plastering, and £10 per square metre for your decorations.

Based on similar schemes, you might allow three thousand pounds for your heating, plumbing, and electrics. All of a sudden, your estimate is starting to take shape. The trick is to break it down into its component parts, and price each one *on its own merits*, taking care to consider the project-specific specifications and the list of materials required.

If you're digging out for the footings, where is that waste going to go? If the answer is 'into skips', then ask yourself how many skips you'll need. Does the project need a full-time site manager?

If so, you need to allow for the cost of this supervision, i.e. someone's wages. How far away is the project? Do you need to allow for travel expenses, and maybe overnight accommodation for your staff if the project is happening some distance away from your base?

Build the project in your head, considering every aspect, and foreseeing any problems.

If there's a large shed at the back of the house, in the position the extension is going to go, then that needs to be moved. If you're going to dismantle it and re-site it, that's a day's work for two men. You need to include it in your costs; otherwise it's going to eat into your profits.

Finally, having considered all of the aspects, and having put costs against each individual item, you can total up your costs and the result will be your estimate.

Personally speaking, I always use a system for double checking my results. Say, for example, your cost model throws out a figure of two thousand pounds for brick and blockwork, based on twenty square metres at £100 per square metre. Assuming your measurements are correct – based on the architect's drawing or measuring out on site – then is two thousand pounds enough to complete the work and allow a little profit?

I always ask myself, who is going to do the work. If it's Billy and his mate (i.e. I know the actual people or the number of people who will be required), then how long is it going to take them? If the answer is one week (remember it's still only your best guess, based on similar schemes and rates of output), then you can double check your estimate by saying 'two men's wages at £500 each for the week is £1,000. Allow five hundred pounds for materials (i.e. the bricks, blocks, mortar, and cavity insulation) and plant (i.e. the cement mixer). That comes to £1,500 for labour and materials.' How much have I got in for the item? Two thousand pounds. That gives me £500 of profit on this item. If it takes Billy and his mate six days, I'm still in profit. Therefore, you know that your price is just about right. Do this wherever possible and you'll have yourself a pretty good estimate, and you'll be a pretty good estimator yourself.

If you win the job and it all goes well, the next time you get asked to price an extension, you'll have *your own schedule of rates* to refer to. This is your best guide, and is better than a Spons or Laxtons any day.

Another important role performed by the QS is that of tendering. This is the art of applying for and trying to win the actual contracts of work. If you're not winning any work, then you and your employer are going out of business. Therefore, you need to tender successfully, and this in itself is a skill.

The first thing to establish is the tender-return date. This will tell you how long you have got to complete the paperwork. Most return dates are rigid. If you don't get your tender back in time, you're going to be disqualified. So, 'how long have I got to complete this' is the first question you should ask yourself. You'll probably be in competition against several other building companies, so you need to remain competitive if you want to win the work, while still making sure there's enough profit in the job should your tender prove successful. It's a fine line, and one that will probably give you palpitations.

The thing to remember is, you never want to win a job and be unhappy that you've won it. There's got to be something in it for you. Don't ever go below a figure that you're happy with, no matter how pushy or persuasive the client gets. This can come with experience, but it's a lesson well worth remembering.

Once you have submitted your tender, you'll want to know that you've been successful, but as soon as you win the job, you'll begin asking yourself *why*. Why was your price the lowest? Did you miss something that all of the other contractors have seen? You'll hope that you haven't; otherwise you're going to end up doing the job at a loss, and your employer is going to have something to say to you that you might not want to hear.

You can ask how your tender sum compared to the others. If you won it by the skin of your teeth, then you've probably done a very good job. If you won it by a wide margin, you've probably made a mistake, or been more competitive that you needed to be. You'll have to ask yourself why, and explain the results to your employer accordingly. Sometimes, if you have direct labour employed with your company, and very little work in the pipeline, you might slash your profit percentage in order to secure some work. After all, you need to do something to keep the lights on. If, however, you're flush with work, you can afford to go in a little higher. If you win the job, there's plenty of profit in it for you. If you don't, it's no biggie because you're busy anyway. These are all things to consider whenever you come to complete a tender.

If you're a contractor, you'll want to be on the tender list with your local authority or with as many people who are dishing out the work in your locality. You may need to do a bit of schmoozing to make sure your company is on those tender lists. If you're a small subcontractor, you'll want to get signed up with the large contractors and private practices in your area who are dishing out the work. A phone call can do it. Send them some literature about your company, or email them a link to your website. Show them examples of your previous work. Find out the name of the person awarding the work and give them a call. Hopefully, you'll have a good reputation with which to impress them. Soon enough, they'll invite you onto their tender list and the rest is up to you. Some companies will have a sales director to do this. If not, do it yourself.

Get your price in *before* the deadline and then wait to hear, or give them a call to find out how you got on. If your tender lost, ask by how much. If you're losing out a lot of the time, maybe your prices are too high. Why are other companies able to do the work cheaper? Is their labour more efficient, or do they have better suppliers who can get the materials at a better rate? You need to find answers to these questions to keep yourself competitive. Like I say, if you're not tendering successfully on a regular basis, you're probably going to go out of business. If you're not tendering successfully, your boss is liable to replace you with someone who has mastered the art more successfully than yourself. It's a fine line, but one that you'll have to learn to walk. It's a skill in itself, and an important one in the role of any Quantity Surveyor.

Learn to tender well, and *always* get feedback from the client. How was my price? How did I do? Was I too high, or too low? Did I miss something out? Did I forget to fill in the questionnaire that came with the tender pack? Did I miss the deadline? All of these things go to make up a successful tender. Learn it all, and do it well, and you will keep yourself and your employer in business. Win a fair proportion of these jobs, and do them at a profit, and you'll probably be flavour of the month in your workplace. The words 'pay-rise' spring to mind. You'll be moving up the ladder to success. Who knows, you may end up as a Senior QS or Commercial Manager or Director. You could be made an Associate of your Private Practice. Ultimately, you'll be a success. And what you will be, above all else, is a good QS.

ESTIMATING AND TENDERING. PART TWO

As a Quantity Surveyor, you will become intimately acquainted with the Bill of Quantities. Sometimes, these are referred to as Bills of Quantity. I don't, to this day, know which is more correct. I suspect they both are, and it doesn't make a whole lot of difference, but I think the former takes precedence, so that's the one we'll use. So, what is a Bill of Quantities?

If you have the time, it makes sense to take that lovely architect's drawing and break it down to its component parts. If you can then quantify each element, it saves time and money for your busy builder, who can then put a price to the quantities contained in the BOQ. This will be the builder's price to carry out the work. Say the architect's drawing equates to 100 linear metres of footings, 60 square metres of floor slab, 300 square metres of brickwork, 450 square metres of roof coverings, etc.

You, or the builder, then apply a rate to each of those quantities, and the multiple of those two sums is your cost for that element of the work. Bite-sized pieces. Like I've said before, we're neither rocket scientists nor geniuses. Bite-sized pieces are good for us all.

If you don't have the time to prepare a BOQ, then an alternative, and one especially suited to smaller jobs, is just to prepare a schedule of work. This is the same breakdown of a project, except you just list the items of work to be carried out, rather than going to the trouble of breaking it down further into actual quantities.

Imagine you're doing a little refurb of Mrs Jones' house. There's not much point in saying 'Brick up a doorway' and 'Create new opening further down the hallway' and then quantifying those items. It is likely to give you a false reading. For example, a doorway is about two metres squared. At a standard £20 per square metre for blockwork, that would give you a total of £40 as a bill item. In reality, you know it might take the lads a whole morning to block that doorway up. It might then take them a whole day to create a new opening. Your materials are £50. The labour is £300. You add your overheads and profit to those sums and the reality is going to be more like £500.

Sometimes, the Schedule of work is a more user-friendly and practical guide to the work taking place. As long as you list every work item that needs doing, then there's no ambiguity and you still arrive at a reasonable price to do the work.

What I've found, in the last five years or so, in these cost-conscious times, is that many tender documents that purport to be Bills of Quantity are really no more than fancy schedules of work. So, they will list out the item, such as 'Brickwork' or 'Blockwork', but then instead of giving a quantity beside it, to which you can apply a rate, and the multiplication of which would give you a cost, they now simply say 'item'.

So, Blockwork = Item, and it's back to the architect's drawing you go. You have to measure it yourself, and give them an all-in rate to do the blockwork. It's a slight cheat. Or a very big cheat. It comes as a result of people wanting to foist the expense of things onto others. Why should they pay their own staff to break the drawing down into quantities when they can just say 'item' and let the builder spend their own time and money doing the quantifying.

So, when you come to either prepare the tender documents yourself, or be responsible for filling them in on behalf of a subcontractor, you will usually be working to either a bill of quantities, a Schedule of Work document, or a BOQ masquerading as an SOW. They're all useful documents. They all do the job. And these will be the tools of your trade.

THE ARCHITECT'S DRAWINGS

If you don't have a Bill of Quantities or a schedule of work, you're going to have to create your own from the architect's drawing. That's if you're lucky. I've priced jobs from a sketch on a piece of paper drawn by the boss, or a client, or a builder. But usually you'll get an architect's drawing. This, too, will be a tool of your trade.

The first thing to say is that you should not be afraid or overawed by this particular document. Essentially, it is a technical drawing of a planned building project. It is a black and white picture of a house, school, stadium, or whatever it is that is being proposed. Within that drawing should be most, if not all, of the elements that you need to price the scheme.

If it lacks detail, or if it raises some pertinent question that you need answering before you can price it, then find the architect's details in the bottom right-hand corner of the drawing (like all art, they are 'signed' in the bottom right-hand corner!'). There, you will find the name of the architect, or their company name, plus contact details.

Give them a call. Ask to speak to whoever it was that did the plans for such-and-such a scheme, and then speak to them as you would any other human being. Explain who you are and what you are trying to do, and ask if they could point you in the right direction with a couple of items that you are struggling to understand. Ninety-nine times out of a hundred, they will cheerfully answer your query, and then you can both get on with your working day.

So, you're faced with what appears to be a blueprint or a complicated technical drawing, and it is your job to interpret that drawing for a builder. You can't just shove it in front of the builder's face and say 'How much is that going to cost to build?' What you need to do is break it down into a language that he or she understands.

You'll need a scale-rule. Quite important, that one. And take a moment to see what it is that the scheme involves. The drawing might show a picture of a large detached house with an extension at the rear. What are you constructing, the house, or just the extension? Get yourself a cup of tea, sit back, and let it soak in.

There'll be various notes dotted around the drawing. Some will be lined up to one side, others will be next to or even within the planned building itself. There might be an arrow pointing to the walls saying 'Facing brickwork' or another arrow pointing to the windows saying 'UPVC double glazed'. They are all there to help and guide you.

Read all of the notes that are relevant to yourself. If you are only providing a price for the roofing works, you don't need to read the notes pertaining to the doors, walls, and windows, etc.

If you've not been provided with a BOQ, you are going to have to do your own take-off, i.e. measurement. Now a very important thing to remember is that architects aren't infallible. They are human beings like ourselves, and are subject to the same daily pressures in work in terms of time and the need to get work out the door as fast as possible. There may well be contradictions in the architect's notes and in the drawing. There might be a note saying 'Tiles on the roof' while in the adjacent commentary it states 'Slates on the roof'. Which is correct? It's probably just an oversight. They make mistakes just like the rest of us. Either pick up the phone and ask them, or maybe even price both options. Just don't be surprised if they don't get it 100 percent right 100 percent of the time.

One thing that you *must* check for is the scale used in the drawing. This is of fundamental importance if you have to extract the quantities yourself. Often, a drawing might state that it is in a scale of 1:100. It should also then state that the 1:100 scale is based on you using an A1 drawing (or A3 or whichever). But what if you're working from home, or either you (or your office) don't have an A1 printer… what use is that 1:100 scale to you now? The answer is, very little. You could look up online how you convert an A1 scale to an A4 or A3 drawing. But how can you trust that this works, or that the architect stated the correct scale for the right sized drawing at the time he wrote it down? You can't, because, like I've said, they're only human, and putting the correct scale on a drawing is *the one thing I've found that architects get right least of all.*

Much better to follow their instructions and see what results you get, and then double check them with a quick check-measure. Doorways are a good one. An internal doorway is usually 900mm wide and 2100mm high. Check it at home on your own internal doors, to the bedroom, living room, kitchen, etc. It's pretty much a universal measure.

So, if by following the architect's instructions, you think you should be using a 1:200 scale to do the take-off for a particular drawing, check it against a doorway. If, using that 1:200 scale, it shows you have doorways that are 4.2 metres high, which would be ridiculous, then you know that the scale is wrong. You try the doorway using the 1:100 scale on your ruler, the doorway is now 2.1 metres high. Seems about right, doesn't it? The scale was wrong. It's not a hanging offence, although it might be if you don't spot it. Do your check-measure to be doubly sure as everything will come from your measure. Get that wrong and your project is in all likelihood doomed, or at least damaged in some way.

If you've been provided with a Bill of Quants, which is effectively someone else's measurements, then at least you have some recourse if those measurements turn out to be incorrect. You can say, you told me it was only $100m^2$ of brickwork. It's actually $200m^2$. Give me some more money (like double the amount).

Best to start small. Say, for example, that the drawing with the detached house and the extension is actually an existing house and a new extension. That's not too difficult to get your head around. You see the house, you imagine how it will look with its new extension, and you can pretty much see what the client – the owners – are trying to achieve. They want a bigger kitchen, or a home-office, and so they are expanding their home a little.

What's it going to need? Well, quite simply, some foundations, some walls, a roof, maybe some windows, possibly French doors. There might be a new kitchen inside, some flooring, some plastering, a radiator, electrical sockets, some lighting.

There's not much more to it than that. And it will all pretty-much be right there on the drawing. From that architect's blueprint, you can measure the individual building elements (or put an allowance for costs against each), and you can pick up on any specific requirements from the individual notes on the drawing; so, if the clients wants a particular type of door or window, or roof, or finish to the brickwork (e.g. render) then it should all be on the drawing.

Work your way methodically across the drawing, taking in the quantities, taking in the notes, and ask yourself if there are any specific points that you should also take into consideration (e.g. It's where!? The Outer Hebrides!). Get as far as you can with it and if you hit a brick wall (not literally, I mean come to a stop) then speak to a colleague, or the architect, or maybe even the client.

The important thing to say is that you have been entrusted with a precious document, the original blueprint for a building project. Someone thought you should have it, because they needed your input, and they needed your expertise. They wanted to know how much it was going to cost to build. And the builder can't usually give a price based just on a drawing, and certainly not for projects that are more complicated than a simple extension. They need help from a QS.

The architect's drawing is one of the most important and exciting tools of your trade. You're going to see a lot of them, and you will grow to understand them more, and to read between the lines, and to overlook or compensate for the occasional architect's mistake the more experienced you get. Don't ever be afraid of the drawings, or the architects. The latter need you as much as you need them.

COSTING AND MEASUREMENT

Costing and measurement are closely related to tendering and estimating. If you're doing one, you're doing some of the other. While I don't intend to talk in too great a detail about the actual act of measurement (there are other works available on the subject which cover it in depth, and I'm attempting more of an overview here) it cannot be overstated that you will stand and fall by the accuracy of your measurements.

There's an old saying, I think derived from the skill of tailoring, that you *measure twice and cut once.* Cloth can be a very expensive commodity; therefore the validity of the statement is evident. I have heard the same expression used in carpentry. It's good advice. I would say that the same thing applies to quantity surveying. *Check your sums.* If you've written 342 instead of 34.2 (as seen in the earlier example) that could produce a huge shortfall in your calculations. If you happened to be talking about strip foundations, at £100 a linear metre, then there is a huge difference in cost between 34.2 linear metres and 342 linear metres. That's thirty thousand pounds of miscalculation right there. By double-checking, again using methods we've already discussed (i.e. you know who's going to do the work and roughly how long it's going to take), then hopefully you'll spot the error in time to correct it. Measure well, using whatever software, scale-ruler, tape measure, or distance-measuring equipment that you have, and then check yourself for errors.

There are square metre rates that you can use as a broad brush-stroke check. An extension to an existing dwelling – assuming it's nothing out of the ordinary – might be £2,000 per square metre overall. Have you broken it down to its components parts, such as foundations, walls, roof, internal plastering, heating, plumbing, electrics, fixtures and fittings, and decorations? If so, how big is the overall area? Take the architect's drawings. How big is the extension? It's five metres by four. That's twenty square metres. Twenty square metres at an industry-wide average of £2,000 per square metre gives a rough cost estimate of forty thousand pounds for the extension. What did your detailed estimate come out at? If it was twenty, or thirty, or fifty, or sixty thousand, there's probably a mistake in there somewhere, or something radically different is going on with this extension, e.g. the gold-taps scenario. You can get these standard square metre rates from a book such as Spons. Sometimes, an Internet search on one of the forums where people talk about having work done on their houses can be highly illuminating.

One word of warning, if you are reading stuff on the Internet, I wouldn't believe everything you read. There's a lot of misinformation and ignorance out there. You have to take everything with a pinch of salt, but if you look closely, read widely, and know what you're talking about, then there are some accurate responses and accurate cost models out there.

There is also something to consider called 'life-cycle costing'. This is basically factoring in the cost of maintaining the building for a certain amount of time. Taking out a 25-year lease? It's your purchase price plus, what, replacing the gutters every five years, or painting the exterior every seven, or re-roofing every fifteen. What about internally? How often do you need to change the carpet, or redecorate? Are the plumbing and electrics in need of an upgrade? Is the lift working? This is life-cycle costing.

When I was studying QS'ing, we had to calculate the centre-line of a brickwork wall and then multiply that by the height before deducting the area of doors and windows to produce extremely accurate measurements. It's important stuff, and you can't go wrong if you have the time and the patience to do it. When I was estimating for that national roofing company, they used to calculate everything down to the last nail. Meanwhile, because they were being so accurate, we were missing dozens of tender deadlines. That made no sense to me. I would just throw an extra fifty quid on the quote to buy us a few bags of nails. That was accurate enough for me. We weren't producing 'exact-imates', we were producing 'estimates'. The clue was in the name, at least that was how I saw it.

Some QS's are accountant types and they will measure everything to two decimal places, and there are a lot of these people about, and they do a good job. Others, like me, take a broad-brush approach and concentrate on adding value elsewhere on the project, in terms of communication, problem-solving, and contributing generally, which is something you can't do too well if you never lift your eyes up from the balance sheet.

The thing is, though, measurement is of *fundamental importance* to what you do. Get it wrong, that's a major cock-up. Get it right, and you're doing no more than you've been expected and trusted to do. You won't get any prizes for getting it right, but you'll sure know about it if you get it wrong. So, measure twice, cut once. In our case, we're not cutting cloth, but once you've written your figure down, the budget will be set, costs will be agreed, and materials will be ordered. You could broad-brush it, or you can take everything to the nth degree, but however you choose to measure, *make sure you measure right.*

THE FIVE DIFFERENT HATS OF A Q.S.

Back in the section on Project Management, we touched upon the five different hats worn by the modern QS. The roles that they have to perform are, principally, Estimating, Procurement, Valuations, Negotiating, and Project Management. It's time to look at each of these specific functions in detail.

Estimating

First up: Estimating. We've already covered this in some detail so I'll keep it brief. Estimating means establishing a reasonable cost for the work that is going to take place. You'll want to be as accurate as possible, so break the work down to its component parts. That way, you'll give yourself the best chance of coming up with an accurate figure and you're less likely to miss something out. If you put in a tender to do some work, and you've missed something out, you may well find that you've just given up all of your profit on the job. So break a job down to its component parts, and price each of those items individually based on the labour, materials, plant, and equipment that you'll need to carry out the task.

I once did some freelance work for an architect who had a very demanding employer. The architect was tasked with doing the estimates in order to quote for work which essentially involved doing up houses for wealthy clients in London. The schemes would typically be worth between one and ten million. Nice work if you can get it. The builder was making a bit of a name for himself, moving between a particular community in North London, networking, and securing one attractive deal on the back of another. Now this architect wasn't really an Estimator. He needed a QS, and he found me. (In truth, I found his son, up a tree, but that's a different story!) Anyway, the architect would draw up the specification, and I would take off the quantities from his drawings and price the work. The builder would give this price to his employer, who would then strike a deal with the wealthy customer.

The architect had once under-priced one element of the work. He had put in an allowance of thirty thousand pounds for the windows. When the work eventually got around to being done, the figure came to sixty, so his employer, the builder, was not best pleased. As he saw it, the architect had under-priced this element of the work and had taken a thirty-grand bite out of his profits. (We'll pick this up, later.)

Due to time constraints, it isn't always possible to obtain accurate quotes from subcontractors, in order to put these figures into your own quotation that will be presented to the client. In that case, you have to 'estimate' it, using rates from previous similar work or based on the best of your knowledge and experience. Hopefully, your figures will be there or thereabouts, and include for an element of profit.

This builder had a particular way of working and, I do believe, it's one of the simplest secrets for success that I've ever come across. The model can be duplicated by you, me, and anyone else, so I'll share this with you now; feel free to get rich on the back of it.

The builder had no overheads. He had no direct labour employed. His talent was in schmoozing his clients and, with each project completed, he also had a body of work that he could point at and say 'this is what we do. We can do this for you too.' As all the homes were impressive, and each worth several million pounds, they couldn't fail to seduce any other prospective homeowners or buyers who wanted something similar, and who could afford the large amounts of money involved.

All the builder would do, after winning the contract, was to *subcontract all of the work* out to other people. He would then project manage the whole affair. His quotation would be made up of all of these subcontractor quotes, and he would apply a twenty percent mark-up to each one. Then, he would tot-up all of these subcontractor quotes, with his twenty percent mark-up on each, and then add a *further* twenty percent to the total to cover his own overheads and profit! And people paid it. He was one rich builder, I can tell you. He was probably doing two or three of these projects a year, and making about half a million pounds in profit from each. That's not bad money, a million, or a million and a half, each year just for overseeing the work of a bunch of tradesmen. Not bad indeed.

Now, to pick up on the earlier story, the builder wasn't too happy that the architect had under-estimated one element of the work, namely the windows, to the tune of thirty grand. The architect apologised and said, ultimately, that he wasn't really an estimator and they would have to employ a QS to do the work in future.

That was where I came in.

When the architect explained the situation to me and told me the windows story, I said that he shouldn't feel too bad about it. By making a thirty grand mistake, he had actually handed the builder a more competitive quote. Who knows, maybe this was what won him the job and gave him the opportunity to make his half a million pound profit on the scheme? The builder, of course, wouldn't see it like that. He would say that he could have made half a million plus the thirty thousand (plus the twenty percent mark-up on that thirty thousand, and the further twenty percent on it when that figure made it into his total!). Maybe he's right. Maybe the client would have paid the sum asked without question, but what if the builder was tendering against other competing builders? Would he still have won the job if he'd have gone in at the higher sum?

Sometimes your estimating mistakes are not the end of the world. Sometimes they let you get your foot in the door. Try not to make deliberate mistakes, but, like I say, it's usually not the end of the world if you do. Anyway, I hope that by including this example here, I've set some of you on your way to becoming millionaires!

Procurement

Another facet of a QS's role is that of Procurement. Basically, this means buying. As the purse-holder of the building project, it is your job to spend the actual budget for the scheme. If you are building your own home, or a five million pound school, or a three hundred million pound hospital, you hold the chequebook and it is your job to spend the budget wisely and ensure that, once the money has been spent, that what you have in front of you is the thing that you set out to build; otherwise, something disastrous has happened.

Say you have one hundred thousand pounds with which to build yourself a house. Now you've spotted a small plot of land for sale, just big enough for your lovely new home. It's twenty thousand pounds, so that's the first bit of money that you have to spend. Now you have just eighty thousand left in your purse/budget. Assuming the land already has planning permission, you can move on to the build phase. If not, you're going to spend another thousand or so from your budget applying for the relevant permissions.

So let's say you now have a piece of land with planning permission and eighty thousand pounds left to build your new house. At this point, you might want to get an architect on board, or a builder with some local know-how. If you're well on your way to being a QS, you may want to project manage it yourself.

You find a structural engineer in the phone book (or ask someone to recommend one) to draw up the technical specifications of the foundations. You get these approved by the Building Control Officer (just call up your local council and book an appointment). Now, you're ready to pour the concrete. Do you want to get a builder in to do all the work, or several separate subcontractors for the groundworks, the walls, the windows, the roof, and the electrics, heating, plumbing, plastering, and painting? Slowly but surely, you'll have to allocate that eighty thousand pounds in order to get the various elements of work carried out.

You might decide to ask three different builders to quote for you. Hopefully, at least one of them will be within your eighty grand budget. And don't just choose the cheapest. Like anything in life, sometimes buying cheap can be a false economy. You have to be sure you're getting the appropriate quality. If you're satisfied with quality, then by all means choose the cheapest, but only if you're sure that you're not buying a dud. So, the budget is yours to spend. Spend it wisely to ensure that your project is fully resourced with labour, materials, plant, and equipment so that your dream comes to fruition. The same is true for any building project. You want the building that you set out to build, on time, and on budget. That is what is known as a success.

Whether you are buying materials for the project, or choosing sub-contractors and tradespeople to carry out the work, it's all buying. It's all procurement.

Valuation

Another component of a QS's role is that of Valuation. If you're working for a large or medium-sized contractor, you'll receive invoices from your subcontractors and you'll have to assess them, i.e. evaluate them. The plasterer says he's finished his work and bills you twenty grand. You check on site and find that they've only done three-quarters of the work, therefore you value the work done at fifteen grand. And so on.

It literally means the value that you place on the work that has been carried out.

You'll also prepare your own interim valuation (e.g. monthly or weekly, whatever), your application for payment, or your final account valuation upon completion of the works. You'll do this by measuring the work you've done up to that point and then by applying the appropriate valuation to achieve your payment. That's valuation.

Negotiation

Negotiating is another part of your job as a Quantity Surveyor. It's a very important part, and it's how you'll really prove your worth to your employer or to yourself. You will be conferring and bargaining on behalf of your client; doing battle, for want of a better word, with the opposing QS, or builder, or sub or main contractor.

Now I'm all for a non-adversarial approach but – at the end of the day – you'll have your budget, and the other person will have theirs, and therein lies the competition. You might find that the other person (or maybe it's you) is being unreasonable and intransigent but – for the good of the project – the key here is compromise. Look at the word. 'Com' is indicative of community, of togetherness. It means something shared. The remainder of the word is 'promise'. Compromise is just a promise shared. Both sides agree to do something.

Remember what I wrote earlier in the book… 'Right is might'! Don't give in to intimidation. You've done the work that you had signed up to do, and you had agreed the price beforehand, therefore you want paying for it. Just don't expect, with all customers, to get an easy ride; stick to your guns. Make them see sense. Try to maintain a good relationship; after all, your paths will probably cross again. Ask for what's right, and then make sure that you get it.

You may need to lower your expectations on occasion. The contractor may have made a mistake somewhere and now they're giving you a hard-luck story. You'll know if they're plausible or not (at least you will with experience). In that case, you can be flexible, and help a client out – but only so much.

- Be honest
- Be open
- Communicate

…then you're negotiating effectively.

Believe me, honesty and communication have never done me any harm in my role as a QS. In fact, they have helped to engender respect, bonhomie, and a good working relationship.

Project Management

Project Management is the fifth and last of what I call the main functions of a Quantity Surveyor. We've already discussed what a bad Project Manager looks like, and we've also seen how much money there is to be earned by doing purely that role (in our prime builder example). It's true that being a Project Manager is a stand-alone position, but as a QS, it's also true that you are partly a Project Manager.

Basically, being a QS is permission to read your boss's emails. It is permission to buy all of the goods. It is permission to have your fingerprints all over the project. If you're not managing the project that you are so inherently steeped in, then you're shying away from the task. You're part Project Manager, like it or not. You may well have a Contracts Manager, or Project Manager, that you're working alongside. In that case, the success of the project depends on you both, and (of course) on a lot of other people and factors besides.

You're a team. And, guess what, if there's no Contracts Manager or Project Manager, that means you're it! Embrace and enjoy the challenge. You'll have a chance to really make a difference to the success of the project. You'll be judged anyway on the financial success of the scheme, so you may as well take ownership, take responsibility, and get it into your DNA.

You're a QS. You wear five hats. Make it happen.

EVERY SITE IS DIFFERENT

I was always taught that *every site is different; you can't price a job without seeing it.* That's pretty good advice.

If it's not possible for you to go there yourself, you need to ask someone close to the job – who knows what they're talking about – if there are any pertinent points to take into account. You may need to qualify your quote, (e.g. to say that you have not allowed for any access issues or the like), unless you've set eyes on the job and you know exactly what conditions you're going to encounter.

I was also told, 'Go to site. It's where stuff happens'. That's a good one, too.

So, while it is possible to come up with generic square-metre rates for most building components, (e.g. brickwork, plasterwork, painting, etc), you need to then tailor these rates to include for the specific conditions affecting this particular project. Where is it? Do you need to allow for travel, accommodation, etc, for the workmen involved? Have you got enough room to erect a scaffold where you need it? If not, do you need to allow for the cost of a crane to lift materials into position? If so, for how long? How many visits? If a crane costs you two thousand pounds a day, you need to know if you're going to need it for one day or five. It all mounts up.

Therefore, if you can, get yourself to the site prior to submitting your quotation or beginning any work. It's where stuff happens, and you don't want to be met by any nasty surprises when you get on site for the build. That could be your profit worn away before you've even begun. Get yourself to site and judge each job on its own merits.

Build the job in your head first of all. Imagine the project unfolding before you, as if you were watching a film. The builders arrive on site, erect their site cabins, put up hoardings to keep the site secure, and then the heavy machinery arrives and starts digging into the earth. Have you seen any problems yet in your mind's eye? Can those machines get in? Do you have room for them? Watch the whole project as if it were a film. If you envisage any problems or setbacks, include the cost of overcoming them within your quote because every site is different.

A city centre scheme may well pose more problems than one in a less-populated location. If the site is on a busy road, think about those poor lorry drivers trying to access the site to drop off deliveries. Can you block the street off for a day when you need to get a crane in to do the heavy lifting? That will need to be coordinated with the council. You may need to speak to the surrounding tenants or businesses who may be affected.

Every project involves a certain amount of juggling, and some involve a lot more than others. Weigh the various factors up at the feasibility stage, or at tender stage, and include any likely costs in your budget. Every scheme is different, and a fundamental part of any scheme is its location. Make sure you know at least a little, if not a lot, about where your project is due to take place before you begin.

Judge each job you come to price on its own merits, because no two projects are the same.

CONTRACTUAL CLAIMS

As mentioned in the Starting Out section, if you pick up a copy of the Yellow Pages and turn to the section on Quantity Surveyors, they will usually list their impressive credentials beside their names and a number will also list LLb, which is a degree in Law.

Why this close relationship between QS's and the Law? Well, it is all to do with the fact that the construction project is based on a contract and, when things go wrong, all eyes turn to the contract.

'Who interprets best, wins,' could well be the motto for the Construction Claims QS. Basically, if you find yourself embroiled in a contract dispute, you'll have to do your research and present a narrative that best reflects your side of the argument. There's my old ally 'communication' again.

Now I'm a great believer in avoiding disputes and trying to resolve disagreements amicably, but nevertheless they do occur. Remember, be prepared to compromise, whilst recalling that right is might.

If you're right, explain to the other party exactly why you're right. If they refuse to see sense, then you may well have to tell the same story to an adjudicator or a litigator or a judge at a later date. If you're in the wrong, no matter how much you try to wriggle out of your responsibilities, you're probably going to have to cough up eventually.

The Wembley Stadium project springs to mind. An Australian company signed up to build the new arena for a fixed sum of £750 million. The eventual cost came in at around £1.1 billion! That's a budget overrun of £350 million. Do you think the contractor just laid down and said 'Gee, we lost £350 million on that one; we'd better go back to building extensions, etc, and try to recover our losses? No way. What they said was 'We lost £350 million. What the hell happened?'

And then the analysis starts. They're looking for answers, and they're looking for cause and effect. They're going to start reading through the correspondence files to try to find some pointers as to what occurred that may have contributed to their loss. They'll sure enough be wedded to the contract too, trying to see if there were any material changes between what was meant to happen and what *actually* happened.

Imagine if the contractor got an email that said 'We're thinking of changing all the seats in the stadium from red to blue. Hold on while we consult with the architect and with some consumer groups.'

There's a four-week delay while they make a decision. Meanwhile, you can't put the roof on until the seats are in, because you can't have anyone working down below while you're lifting the steel trusses into place, etc. So now you have a four-week delay, but that has pushed the roof back into winter, and you can't lift those steels in the heavy winds, so now you have a three-month delay.

Assuming your preliminary costs – staff, welfare facilities, security, site cabins, hire of equipment, etc – are running at £250,000 per week, then a three-month delay is over £3 million. If you can make a case for that and a hundred other things, then you can legitimately start to claw back a significant portion of the £350 million that you lost.

That's the Claims Consultant's role for you. It's a growing field within construction and quantity surveying. Remember, gather your facts, and put together a narrative. Try to come from a good place. Very few of us want to be lying on the witness stand (or in front of whoever it is that we're explaining the situation to).

In the Small Contractor section, I quoted Gandhi. He said, if you're in the right, nine times out of ten, the law will come to your aid. The cases you lose are typically the ones where you didn't do your research properly or you didn't get your narrative across.

If I had my time over again, I would probably study Construction Law and specialise in that. Certainly, Construction Claims seems to be a legitimate area of our industry now. It may appeal to you, and I can only wish you good luck and say I'm slightly envious of your choice. One degree is enough, though, whether that's QS'ing, Construction Law, Project Management, Building Surveying, Architectural Technician, Construction

Management, Planning Officer, or any of the myriad roles that could make you a construction professional. And even if you're just a regular QS, you'll still find yourself doing a bit of Claims work. We're multi-skilled, multi-functioning, multi-faceted QS's.

Claims Consultants, especially those allied to the RICS, can earn hundreds of pounds an hour. If you could build up a client list or advertise your services effectively enough, you could probably hire yourself out as a mediator or expert witness, or just someone that two warring parties agree to appoint to try to sort out their differences.

You'll often find that there are three sides to every story. Your side, my side, and the truth. It's never usually black and white. There might be some middle ground that both sides can agree on.

Often, as we saw earlier with the window installer and the bad project manager, these issues get settled before they ever reach court. They get contentious without actually becoming contractual.

Quantity Surveyors are very much sought after for this kind of work, both for their knowledge of construction and for their negotiating skills. In your day to day role, you will be sorting out small disputes. You might be chasing a client for non-payment of an invoice. Why haven't they paid? Are they unhappy with your work? Has your company not fulfilled its brief?

Get to the bottom of it. Sort it out. That's your job. You're the QS.

NEGOTIATION

Gandhi (how I love that guy) once called negotiation, 'The beauty of compromise'. Eventually, I understood his point. 'Com' and 'promise'. A promise shared.

Another expression I like is 'Never try to overkill a good opponent.' I think this should be true in any negotiation. Take as much as you can, but don't try to destroy the other party. You're probably going to meet them again somewhere down the line. You were supposed to be partners, after all. Remember *the joy of compromise.*

In any negotiation, you need to be prepared. First and foremost, *know your costs.* That way, you'll know how much you've got to work with, and how much – if anything – you can afford to give away. You should aim to have three positions before you go into any bargaining meeting. This could equally apply to your own salary negotiations.

Your first position is the figure that you hope to achieve. If you have a great day, and the other side acquiesces to your every whim and desire, then your first position is the figure that you would be delighted to accept.

The second figure is the position that you're in all likelihood going to achieve. You state your case for the first figure, but then the other side states their case for a considerably lower sum. They have some valid points to make, therefore you concede a bit of ground and you end up at your second figure. That's the compromise position, and it's the one where two good negotiators are probably going to end up. Call it a draw.

The third position is your absolute bottom-line figure. If you go into the negotiation all tongue-tied, and the other side wins every argument, then you should have a figure in mind that is your absolute rock-bottom sum, below which you just won't go, no matter how hard you're pushed. It's worth knowing this figure – your third and last position – before you enter into any negotiation, otherwise you may end up settling for even less than this amount, and you'll have a financial disaster on your hands.

So, aim for three positions: your dream result, your compromise sum, and your bottom-line figure, below which you'll just have to walk away and leave it in the hands of the arbitrators, adjudicators or litigators. Once the other side can see how serious you are, and once you've explained honestly and openly the reasons why your bottom-line is what it is, they'll probably come around anyway as they should be able to see the sense in your position and they shouldn't be trying to 'overkill' you.

I remember reading about some Hollywood executives who were in negotiations to buy a rival studio. One side arrived with a briefcase in which there were a series of envelopes, each with increased offers. I think there were about five envelopes. When the other side caught sight of the briefcase and the envelopes, and rightly guessed what they meant, they turned to the other side and said 'don't even bother with envelopes one and two, we'll start at envelope number three.' That's a pretty good opening gambit. They probably settled for envelope number four. Overall, achieving four out of five is a very good result. Logic tells us that if both sides negotiated as well as each other, they'd have probably come out at envelope number three, the mid-point. Four out of five is good negotiation. (I guess this story also tells us not to let the opposition see your envelopes, metaphorical or not!) Try not to show your hand.

When I was working as a Claims Consultant, my friend and mentor told me about one negotiation that he was involved in. The sum at stake was an overly-exaggerated million pounds. Basically, my colleague had over-inflated his figure to give himself some wriggle-room. He never expected to get it, and didn't even think that he deserved that amount, but he was exaggerating in the expectation that he would get a significant portion of it. He actually only thought he was entitled to about a quarter of that sum. When the other side came to the bargaining table, they opened up by saying that they couldn't afford to pay a million pounds. They said they only had six hundred thousand pounds available, and that was it. My colleague immediately accepted their offer, saying 'well if that's all you've got, there's no point in me trying to get any more.' Secretly, he was delighted. You could say that the other side negotiated badly by putting their cards on the table straight away I would agree. Try to be a good negotiator yourself. It's a big part of your job, so aim to do it well.

FREELANCING. PART TWO

I had freelanced for four years earlier in my career, then found a port in a storm when the next recession hit. I spent four years in an estimating department for a national roofing company, and then I went QS'ing again.

I subsequently spent two years working for a roofing and cladding company, and left there when it became apparent that I was going to be working away from home for much if not all of the year. I just couldn't do it.

I then spent six months working for a family-owned building contractor and developer. I loved the work. It was varied, with a bit of everything, from working with some big utility company clients, down to doing an extension at the back of Mrs Jones' house. I really loved it. And the offices were close to my home. My colleagues were great. The work was great, the environment was brilliant, and I liked my colleagues. So why just six months?

Being family-owned, the directors were control-freaks. I kind of thought I knew what I was doing at that point in my QS career. My results spoke for themselves. All of my jobs were making money. Usually lots of it. But my boss would give me a look, every time I left the office to go to site, as if I was heading off to the bookies! He wanted me sat where he could see me, and that way he would know that I was working! He'd advise that I just give the Contract Manager a few words of advice at the start of a job and then leave them to it. None of this 'get it into your DNA' stuff.

Finally, I gave Mrs Jones a quote for a refurb of her small terraced home. Mine was the highest of three tenders she had received but, (communication again) I was the one that she felt she could most trust, the one who would hold her hand through the project, answer the phone when she called, and make sure that she got what she wanted.

My boss wanted to know how I had acquired the figures that had gone into my estimate that made up my tender. I told him that I had broken the work down into about forty different work items, such as heating, plumbing, electrics, plastering, painting, and the like. Where possible, I had broken those items down further into quantities by measuring each room (roughly – not to three decimal places) to get 150m^2 of plastering, etc. I'd then applied a reasonable m^2 rate to these items (I know what you're thinking, you're basically just doing the job of a QS. You're right). Where I couldn't quantify an item, say for the electrics, I'd either phone one of our subbies and get a ball-park figure, or I might say to myself that the last house of similar size and specification came out to three grand, so I'll put that amount in again. It all added up to a figure.

I knew – because the client (Mrs Jones) had told me – that my price was slightly higher than the other two quotes that she had received, so it wasn't a bad bit of estimating. I certainly hadn't undervalued the work. If anything, it was overvalued, because it was the highest of the three. It wasn't wildly over-inflated. It was there or thereabouts. It was pretty good estimating to be fair. And, to top it all off, it was still the price she was willing to accept because I had sold her the reputation of the company and I had spoken to her in a language that she understood and trusted. And she was right to, because I would have fought tooth and nail with everyone involved, including our own to make sure that she got what she wanted, which was the job that she was paying for at the price that we had agreed. Seems fair, doesn't it?

Guess what my boss thought? That I was playing the stock market with his money. That the figures contained in my estimate weren't actual figures. They were a guess! Yes, I could add. That's what an estimate is. But it's an educated guess, by someone paid to do just that. If you don't want someone who is prepared to estimate, don't hire a bloody estimator or QS!

Do you know what he wanted me to do? Put twenty subcontractors in a minibus and head around to Mrs Jones's house. They would all weigh up what was needed and give exact quotes for the work. That would be the price we would put forward at tender.

Well, what if Mrs Jones didn't want 20 blokes in her house? What if those 20 blokes were busy with their own work and therefore unavailable? What if they just thought I was crazy for even asking and stopped taking my phone calls? I would never get another quote out the door. Doesn't 20 years of experience count for anything? Doesn't the fact I'd smashed it on every previous project earn me a little license to guesstimate? I'm prepared to take the rollicking if I've cocked up, but if I haven't, then I've done my job correctly using the time and resources available.

For six months, we locked horns over how to be a QS. One week, I'd had enough of his nonsense and walked out. The senior QS talked me into coming back. Two weeks after that, three days before my wedding, the boss had finally had enough of my stubborn ways and sacked me. My six month probation period was up. They weren't going to offer me a permanent position. The way I saw it, I should have got a medal for lasting that long. And, in fact, he was the inspiration for this book. I suppose I should thank him for that. I thought 'This is how to be a Quantity Surveyor.' I wrote this book for him.

Then I went on my Honeymoon, and then I came back and got a job. I stayed there for a year, and then my wife was about to open up a shop. I knew she needed a bit of support. And so I left my full-time job and I went Freelancing once more. It was a year where I learned a great deal, trying my hand at areas of construction that I had no previous experience in. I did it for a year.

I call it the year of getting sacked!

THE YEAR OF GETTING SACKED

I got a temp freelance position with a Painting and Decorating contractor. This was a medium-sized contractor doing work for a lot of large contractors. The overall schemes were generally quite big, and cost millions of pounds. Our order values ranged from maybe £50,000 to £500,000, and we were working on at least half a dozen schemes at any one time.

Do you know what I loved about this position? Well, it was based in my home city, so travel wasn't an issue when it came to getting to the office. The projects themselves were based within about a 50-mile radius, so trips to site were usually an hour's travel-time at most. But, apart from the location, the thing I most enjoyed was seeing construction in progress.

When I worked in an estimating office, I never once visited a building site in four years. It was like I was employed in a completely different industry. Even when I worked in private practice, trips to the site would be rare, maybe once a fortnight. When I managed the maintenance programme for a large property portfolio, again I never set foot out on site. And that's a shame. Because building is what our game is all about.

The construction industry is a dynamic workplace. Empty plots of land are transformed into (sometimes) magnificent buildings. There is change. There is evolution. And it happens in stages. If you were to watch a time-lapse recording of a building project, it would make for quite exciting viewing, I think. The transformation would be quite dramatic. A brand new building rising out of nothing.

The joy of being a QS for a painting and decorating contractor is that you arrive on site fairly late in the process. As far as building packages go, the decorations are one of the last things to do, so it was interesting to see the building in its almost completed state. Usually, projects are running late, there is pressure from the client to get the project over the finish line (and the building handed over), so there would be lots of different contractors still on site trying to get the thing finished.

And there we would be, in the midst of it all, trying to make it look pretty. Once it's decorated, usually the only thing left to do is the external landscaping and then you're done. So, it was great in terms of improving my building knowledge – seeing construction taking place before my eyes. It was also nice to be adding the finishing touches to the scheme.

Then, after a month, my employer found a young graduate QS that he could take under his wing and train up for a lot less than he was paying me. Just like that, my contract, which was running on a week-by-week basis, wasn't going to be renewed. I was a bit upset, to be honest, because I liked the company and I liked the work. But I was freelancing. There was nothing I could do about it. And that was sack number one.

I then found a short-term contract with a company doing an internal office fit-out in a well-known office building in the heart of the city where I live. It was a bustling complex. Hundreds, if not thousands of office workers pouring in and out of the building every day. And there we were, taking it over, one floor at a time, knocking internal walls down and then rebuilding them, introducing new floors and ceilings, changing the layout, upgrading the electrics and data points and making it fit for the 21st century. The tired old office building was being given a make-over. We had demolition crews on board, joiners, a plasterboard partition company, plasterers, painters, labourers, plumbers, and electricians. It was a £3 million pound refurb. I got three months out of it. I was let go when my probationary period ended. They didn't think I was worth the (not bad) money that they were paying me. I couldn't really disagree. In the interim, my long-held dream of being a writer was coming to fruition. After 20 years of trying, I secured two book publishing deals within the space of a month. You could say I was a little bit distracted.

Anyway, I got three months out of it and – as I was freelancing anyway – no one was going to question the duration of any one contract. I could come and go as I pleased. And I did, but you could call that sack number two. Of course, I didn't like to see it like that. I was freelance, so essentially I was working for myself. The only person who could sack me was me. I'd lost a client, that's all it was, and I found the next one the very next day.

Sometimes, when I've been freelancing and able to take a job at short notice, I've had employment agencies tell me that I am the only available freelancer on their books. That's not a bad place to be in the job market, occupying a field of one. Who they gonna call? It ain't Ghostbusters, that's for sure. They're gonna call you.

The next place I rocked up at was a two-day-a-week gig for a property developer. I'm going to talk more about property developers (and the scheme I worked on) in a little while, because this is a growing section of the construction market, and I think that developers are little understood. It was certainly an eye-opening experience for me. For reasons to be stated in a later case-study, I'll keep my powder dry for the moment. Suffice it to say, after a couple of months came sack number three. Or should I say, I lost another client!

Then came another part-time two-day-a-week stint with yet another developer (and yet another case-study for us later), which led to the loss of client number four. Okay, sack number four.

Along the way, while all of this work was going on, and while I was writing two books on the other five days a week when I wasn't part-time QS'ing, I had a couple of private clients who needed a bit of Quantity Surveying support. These clients included a groundworks contractor bidding for local authority work.

I would drive over to their office in North Wales and spend the day preparing estimates and filling in the tender documentation. I'd then sit down with the MD and give him the chance to check my figures before driving the tender letter to the council or architect's office and submitting the tender.

I also priced a couple of jobs for a small building contractor. The daughter of the owner had spotted an advert I'd placed in a local newspaper for QS or Estimating services. She knew that her dad was looking for some estimating support. She saw my ad, gave her dad my number, and I was soon in their offices getting £500 a quote.

I even ended up doing a couple of days a week for the roofing contractor I had left when I'd decided to go freelance. We stayed in touch, and I even got him to tender for a couple of roofing packages for schemes that I was working on; one he won and one he didn't. When his own estimator, my previous replacement, decided to head off on a six-week holiday, I stepped in to help him out, doing two or three days a week. When his employee failed to return from his holiday, and as I had just experienced the heartbreak of sack number five, I asked for my old job back.

And thus ended the year of getting sacked.

Throughout those 12 months, I did many things in the industry that I had never done before, that I had no construction experience of. It was like four years of work experience crammed into twelve months. I had never worked in new-build housing, nor had I done an internal fit-out, and I had never worked for a developer. Now I had experience of all three. So, freelancing: good for earning, good for learning. I highly recommend it.

Just try not to get sacked!

THE PROJECT FROM HELL!

You might be a good QS. You might have a good team around you. You might think that you know what you're doing, but you can still find yourself in a mess. It's what you get paid for, dealing with problems as they arise. And they will.

I once said to a Contracts Manager, why do we spend all our time dealing with problems? His response was, 'Well, if all these jobs were running smoothly, we'd all be down the pub!' So, problems will arise, and it's our job to sort them out. We're managers, after all.

We will meet challenges on the way in any given project. Occasionally, though, (and fingers crossed only occasionally) you'll find yourself embroiled in a job from hell. One of my own came about like this.

I got a phone call. So far, nothing to worry about. A client said he had received a quote from ourselves, and he was interested in placing an order. So far, so good.

I asked for the quote reference number in order to dig out the file. The number he gave me surprised me. It was more than a year old, and done by my predecessor in the role. I'd have to look into it, I said, as the quote was out of date. I needed to see if we could still live with the price we had given him. And as I hadn't done the quote, I needed to get my head around it and see if it was something that I could live with. And then I added, somewhat presciently, 'You've sat on this quote for a year. I bet you're going to tell me that you want us there next week?' He laughed, and said yes. It would be the last time that either of us would laugh for a very long time.

I found the old quote. The job was a small one. It was only worth £7,500. It included costs for us to provide the scaffold, etc, on top of the work that would be carried out – an overlay to an existing roof and a valley gutter. I said that if he'd cover the prelim costs, (e.g. provide the scaffold), we'd stick to our original price. He agreed.

When could we start, he wanted to know. Well, as soon as you give me an official order, I replied. I'm not going to spend two grand on materials until I have it in writing that this deal is going ahead.

Straightaway, we were under pressure. When are your men and the materials coming to site?

Hold on a minute; you only gave me the order two days ago. The ball started rolling then, and things take time to happen, like ordering the materials and allowing the suppliers to make up the batch and get them delivered. We'll be there next week. Already I was thinking 'what have I gotten myself into?'

Then their contract manager changed the scope of the works. We should have been in and out in a week, two at the most. He didn't like what we were planning to do (which was their scope of works!) and he wanted to do something else. I didn't even charge him for the change. I should have done, it was a variation, but it shouldn't have been more than another day or two's work, and I was trying to roll with the punches.

A day before we were due to finish the revised scheme, the contracts manager realised that he shouldn't have actually been doing *either* of the two schemes he'd enrolled us on already. He was meant to be doing something else entirely. We'd stripped out the slates on the roof at the side of the valley gutter, in order to accommodate scheme number two, the scheme he'd revised halfway through. Now, he wanted an elaborate tapered insulation system to go into these valley gutters. They would take weeks to manufacture, and weeks to install. My little 'in and out' in a week job was growing and growing. Often, that's not a bad thing, but we were flat out elsewhere, and I'd only accepted the job because it was a little scheme and it would help them out. I should have known better.

And because, when we eventually came to install the tapered insulation scheme (or their scheme number 3), we would need to remove the very same slates that we were about to reinstate, the site manager wouldn't allow us to do so, fearing he would be charged again when we put them back. We put a temporary Visqueen in, pointing out that it might last overnight but it wouldn't give a month's worth of protection, which was what they needed.

It didn't matter. They didn't want to pay for us to re-slate it (even though I wasn't charging them) when they would have to be stripped out again. And boy did it rain. And, of course, it leaked. It was a school building. It was very much a live project. The headmistress was none too pleased. The contractor took aim at us and said they were going to charge us for the leaks. The cost? £7,500.

I responded with an email. Don't ask us to take the roof off, stop us putting it back on, and then complain that you've got leaks!

The revised insulation scheme cost another seven grand. I invoiced ten grand for work carried out. They valued our work at £100. Now, you can expect a bit of to-ing and fro-ing with your invoices. Not everyone agrees. Things get re-evaluated. They might knock ten percent of your valuation. But from £10,000 down to £100? That was a new one for me.

I was straight on the phone. They backed down. They gave me another grand!

No matter what I tried, they laid the blame squarely at our door. The new scheme, scheme 3, was eventually delivered to site. Guess what? That one didn't work either. We were on to scheme 4. They were making it up as they were going along, and we had our wagons hitched to their train. And when all was said and done, they just turned around and laid the blame squarely at our door.

We were a good scapegoat. The roof's leaking. Blame the roofer. Obviously. I received a call from the surveyor at the local authority. They were ultimately the client, acting on behalf of the school. He phoned me up to give me a bollocking. I headed him off at the pass. I knew what scheme he was going to ask me about. I told him my tale of woe. At the end of our conversation, he said, 'three-quarters of what you've just said, I haven't heard before.' Someone was obviously being very choice with their words, and was only giving him half (or a quarter) of the story.

I can't imagine how many nights I woke up with nightmares about this project. At the time of writing, almost twelve months later, we are still trying to get paid for the work we've done.

In hindsight, we should have carried out the initial works and then left them to it. You try to do your best, you try to help them achieve their vision, but if they don't know what they're doing, you're really going to hell in a handcart. And paying for the ride.

Try to step back from the mess. Try to keep your head when – all around you – people are losing theirs. It's easier said than done. I know, because this job just ran away from us. It can happen. Try not to let it happen to you.

COMMUNICATION

I get job alerts all the time. I got two the other day. In the candidate profile for the person that they were looking for, both of them stated as their number one priority, 'Excellent communication skills'.

As mentioned several times before in this book, I believe that the role of a Quantity Surveyor is fifty percent communication. If you can't convey information clearly and accurately, you're more likely than not going to run into trouble. A well-articulated email or phone-call, or contribution at a meeting, can stop problems occurring, or can get you out of a sticky situation should one arise.

Winners have a goal and a focus. Your project has a goal and a focus. If you can't state unequivocally what that is, you and the rest of the team are going to end up floundering in a sea of indecision.

Now, not everybody is a natural-born communicator. Still, if you're a Quantity Surveyor, or any sort of professional person, I'm guessing you are blessed with a fair amount of intelligence. Therefore, you should be able to communicate. If you're lacking in this department, as a professional person, I can only surmise that you're either unwilling in this respect, or you're not giving this aspect the importance that I'm convinced it deserves.

To me, it seems obvious, and even employers are now stating this as their number one priority – even above qualifications and experience – so I guess that I'm not alone in valuing this core skill.

I would break it down like this. The role of a QS is ten percent qualifications, forty percent experience, and the other fifty percent is spent communicating that experience. The qualifications merely get your foot in the door. Your ability is largely contained in your experience and the ability to communicate that experience.

Explaining things simply is a skill in itself. Some people seem to pride themselves on their ability *not* to be understood, as if that somehow makes them the smartest person in the room. If no-one can understand them, that must make them really intelligent! I couldn't disagree more, and usually it means that the culprit is trying to cover up for a multitude of deficiencies, either in themselves or because something has gone seriously wrong with the project. That person can usually be found running for the hills when eventually the truth is discovered, which it will be.

There is absolutely nothing wrong with saying 'good morning' to someone. That sets you up for the day, and it's communication. There's nothing wrong with saying 'when are the plasterers due on site?' or 'when is the crane arriving?' If you get an answer that doesn't fit with your expectations, you may have a problem. Talk to the architect, talk to the buyer, talk to the supplier, talk to the project manager, talk to the lads and girls on site.

If you've identified a problem, sort it out. And the key to heading off any problem is in communicating – to all the stakeholders – what's going on and putting the appropriate remedies in place.

Not everyone will need to know everything. That's a judgement call that comes with experience, but don't keep people in the dark who should know what is going on; otherwise it will be *you* who is running for the hills. There is great humility in making yourself understood, and your project, and every stakeholder involved in your project will benefit as a result. So try to be clear, try to be concise where possible, and try to keep everyone in the loop by being an effective communicator.

Employers everywhere seem to value this more and more, and I, for one, am in total agreement.

PAYMENT PROBLEMS

If you're a QS, your job is primarily concerned with numbers, and what those numbers represent… which is money. And what you will find is that those who have it, don't like spending it. In other words, if you're the person asking for the money, you usually end up jumping through hoops and almost pulling your hair out in order to get it. In fact, in a previous role, I said on several occasions that I was going to change my job title from Quantity Surveyor to Debt Collector, because that's what I spent half of my time doing.

Assuming that the main contractor is being paid by their client – whether that be a private individual, a corporation, an organisation, or even a local or government council – then once they are in receipt of those funds, they are sitting at the top of the money tree. What follows is an exasperating run-around where the smaller subcontractors are made to wait patiently for whatever scraps the main contractor deems fit to pay, at a time of their choosing. Never mind that you have a contract. Never mind that the contract will contain payment dates or payment terms. It's all hogwash. You will get what you're given (if you're lucky) at a time of the main contractor's choosing.

You might not believe me, but I can 100% guarantee that it will happen. Clients, contractors, whoever, for the most part, would rather change their names and run off to Timbuktu than pay you for the work that you've done.

The way it should work is this. Say you're going to be on site for six months building the walls for a new housing development. The plan, at the outset, is that you will invoice on a monthly basis, usually at the end of a calendar month. The contractor then has thirty days in which to pay (if thirty days are the payment terms). It could be more. It could even be less than thirty days, though this is less usual.

Now, in order for the main contractor to manage his cash-flow and draw down their own money from the client in a timely fashion, they ask that you submit your invoice about a week before the end of the month. This gives them a week to check it before they put it on their next payment run. So far, so good.

If you miss that date for any reason, you may find you've lost another month, even if you were only a day or two late. You may put your invoice in on time (important that you do this), and then you wait patiently for the money to hit your account a month later.

And then it doesn't. You phone up. 'Where's my money?' you ask.

'Apologies. It was missed off the payment run', or 'the commercial manager is on holiday and hasn't yet approved it,' or 'it's sitting with the financial manager who signs all the cheques.' Basically, it's in the queue (right at the back), and so are you.

What happened to your thirty-day payment terms? I'll tell you what. Reality happened. They'll pay you when they want to.

And that's the half-decent ones. The good ones, of course, will pay you more or less on time, but I'd say they make up, on a good day, about half of the people that you will deal with. Another quarter will consistently pay you late, sometimes very, very late. The other quarter will make you hop on one leg for your money, and then tell you that they didn't mean that leg, they meant the other one. And now stand on your head, while whistling a happy tune. Okay, they'll eventually say. Sorry for messing you about. We had a few issues with our funders. We'll pay you next Tuesday. Promise.

And guess what happens then? They'll make you do it all over again as Tuesday comes and goes without any money landing in your bank account. It could be six months, if ever, before you get your cash, and then sometimes only a fraction of it.

It will drive you insane. You'll boil over with the injustice of it all. But it will happen.

What can you do about it? Well, if it's a particular person that is being obstructive, try and get around that obstacle. Do they have a boss? Is there someone in their accounts department who might let you bend their ear and who might listen to reason? Can you reach the commercial manager or the finance director?

The best thing you can do, without doubt, is monitor your projects. If you've got a six-month contract, and the first payment hasn't arrived, give them a seven-day notice that you're going to be leaving the site and that no more materials will be delivered until you get some money from them. That often does the trick.

Forget that you're talking business, talk practical issues with the contractor. Communication again.

'Look, we are right at the top of our credit-limit with our suppliers. I understand you've got issues with the funders (or whoever – *you're talking crap anyway*) but I can't get any more gear on site because I haven't been able to pay my bill with my suppliers because you haven't paid me. Obviously, if I can't get any materials on site, there's no point me sending our labour to the site because there won't be anything for them to install when they get there. So, pay us, and we'll come back.'

They can't really argue with that. Of course, at some point, you won't be able to use that argument anymore because you'll have completed all the work and will still be waiting for the final invoice to be paid. In that case, try and manage the risk as best you can by maybe leaving a few minor work items outstanding as you seek to recover the larger portion of the monies owed.

What else can you do? Well, you're a QS, with negotiation and communication skills. Bring all of that to bear. Try to get your tight-fisted late-paying client or contractor to see some sense. You're not there to fund their jobs. You're there to do a job for an agreed amount of money. When that doesn't happen, draw a line in the sand, and then bug the life out of them until they pay you. Get inventive. Get creative. Some people have even been known to get nasty. Whatever works, make sure you get paid, so that you can stop being a debt collector and get back to being a QS.

MANAGEMENT

If you're a Quantity Surveyor, you are part of the management team. If you're a senior QS, or if you're a QS working for a small company, you'll be managing a team of your own. Can you delegate, motivate, and manage? It will be expected of you, and it's something that you'll have to get used to in your role as a construction industry professional. The thing is, you're one of the lucky ones. I started out in the construction industry as a labourer, and then became a roofer's mate, doing the heavy lifting.

Eventually, I realised that I wasn't getting anywhere, didn't have a trade as such, and was better suited to using my head instead of my hands. You'll be working with others you might consider less fortunate than yourself. Some of them will quite happily be working on the tools, as painters, bricklayers, electricians, plumbers and the like, but some of them will be drawing low wages at the unglamorous end of the business.

They'll be working on your projects. Can you talk to them? Can you listen? Can you take their comments on board and respond to their concerns, or get them the materials and the other bits that they need to go about their work? If you're a good manager, you'll be able to. These are the people at the coal-face. No-one knows better than them the strengths and weaknesses of all your decisions. You'll want to keep them onside in order to motivate them; otherwise your only option will be to bully them into doing what you want.

Treat them with respect, and they'll go the extra mile and more for you. Treat them like dirt, and you'll be on your own, and your projects will suffer.

And, like I say, they are the ones who know what's really happening on site. If you're out of the loop, your project will be hurting. You won't be effectively communicating if you're not talking to everybody, which includes the people on site doing the actual work, and not just the rest of the management team, the client, and the architect.

It's not hard, management. It brings its own pressures, but also its own rewards. Consider yourself lucky that you're at the management end and not at the dirty end of the work. But make your decisions good ones, or you'll lose the respect of everyone whose job it is to respond to, and carry out, your instructions. Always remember, you're a team, and though you may be the manager, you're still a part of it, and a vitally important one at that.

If you like football, or any sports, then you'll know what a manager is. There are good ones and bad ones, and even the good ones sometimes come unstuck because they don't have the people skills, or the motivational element, or the compassion, to get each cog in the wheel moving in the right direction.

Some people respond to an arm around the shoulder, and some will respond to a kick up the backside. As a professional, I've never really responded well to a rollicking because I think I'm working hard, doing my best, and that I have something to offer. I've usually found a telling-off to be de-motivating, unless I believe that I've deserved it, in which case I'm prepared to pull up my socks. You're a manager, much like a football manager, now go win the league or a cup or, given that our job is neither that glamorous nor that well-paid, go achieve a winning project. That's your success.

As a former labourer, having done some of the worst and worst-paid jobs imaginable, it was life-changing to move over to the white-collar side of the industry. Yes, I gave up three years of potential wages to study the qualifications that I needed, but I was on a mission, and it's one of the best decisions I've ever made.

I realised, early on, that I might write an item on a schedule of work or on a Bill of Quants that said 'Dig a pit two meters square by one metre deep.' Guess what? Back in the day, I'd have had to pick up the spade and dig that hole myself. Becoming a professional meant someone else would have to dig that hole (sorry, whoever you are). I've already been there and done that. Maybe you never have. Hopefully, you'll never have to. Enjoy being a manager. It certainly beats digging holes.

CASE STUDY 1. M.O.D. LYNEHAM

I hope I'm not breaking the Official Secrets Acts here, but I'd like to talk about a large infrastructure scheme that I worked on recently. The project was a £168 million pound project to take the former RAF airbase at Lyneham in Wiltshire and turn it into the new headquarters for the British army, navy, and air force.

Two of the largest construction companies in the country teamed up to manage the project together; such was the scale of the enterprise. This is what is known as a joint venture.

The scheme was a mixture of new-build and refurbishment. This tired old airbase was going to be given a new lease of life. There would be new accommodation blocks for all of the troops, as well as a new sports block, medical facilities, bar, canteen, and all of the industrial-scale buildings that they would need to train, learn, and work in.

So the Government gave the two contractors the £168 million and said 'crack on and give us a new state of the art facility for the Ministry of Defence.' They had a shopping list of what they wanted. The contractors were going to deliver everything on the list for the contract sum. But they couldn't do it alone. They were literally going to hand out sub-contract packages worth millions and millions of pounds. There would be a mechanical and electrical package, a groundworks package, a bricklaying package, and so on, and so forth.

Now, the QS for the main contractor (there were actually a lot of them, each looking after four or five separate packages) gets three quotes in from subcontractors for each parcel of work. He will only select the lowest of these once he is satisfied that all three tenders are of comparable quality. He may get a price that he's happy with, but the subbie has just won another big contract elsewhere, so do they have the resources to do the job? Maybe not. In that instance, the QS might go to quote number two. He might even use the fact that they're not the lowest as a bargaining tool in his negotiations with subcontractor number two.

He might say, 'look, I'd like to give you the job, but you weren't the lowest. I need you to price-match the lowest quote, or maybe reduce your quote by, say, five percent and I'll give you the job.'

The subbie looks at his price and thinks, if I knock five percent off, I've still got a decent margin. The quote is reduced and the subbie gets the job.

At Lyneham, if there were ten QS's in the team, then each was effectively spending nearly £17 million of the government's money. This money was being spent on literally dozens and dozens of buildings across this vast military complex.

The estate was enormous. There were aircraft hangers, tank repair shops, huge cavernous buildings that served some unknown purpose, but which might now be scheduled to be brought back to life to serve their time anew.

At the time, I was working for a roofing and cladding contractor. We had an initial order of just over a million pounds. This eventually grew to over £4 million. I was the project QS. As the initial contract had been priced by our Estimating department, my role was merely to carry out our monthly application for payment, checking work done against the bill of quantities and asking for our money.

Due to the ever-expanding nature of the project, and our own role in it, what started out as an initial monthly site visit to do my valuation, eventually became a weekly visit, and then I moved there full time. It literally just grew and grew, and it needed the constant attention of a QS.

The place was immense, and there was an army of people working there, even before the real army was due to come to stay. The site was awash with activity. People were almost falling over themselves, trying to get their work done, working around other contractors, sometimes with conflicting agendas, and always under the watchful eye of the main contractors and their management team.

The main contractors took real pride in their prestigious project, and woe–betide anyone who wasn't pulling their weight or meeting their targets.

One quirk of the scheme was that each of the buildings being worked on had its own individual project manager. Just as the

QS's had been assigned a handful of packages to look after, the project manager's each had a building or a couple of buildings for which they were responsible. Because the whole project was so big, it made these mini-projects seem quite miniscule in comparison. I wondered why each building had its own manager. But then it dawned on me that these mini-schemes were really not so small.

For example, the new medical building cost about £3 million. A scheme that big in the real world would require its own project manager. Of course it would. So that's what they did. We were working across a total of about fifty different buildings on the site. That meant that I had effectively fifty different clients. That would be a heavy workload at the best of times. Here, the only difference was that they were all in one location: RAF Lyneham.

You couldn't go anywhere without bumping into a Project Manager. It was incredibly demanding, but quite invigorating too. Each subcontractor had been allocated an office in a large disused building in a quiet corner of the site. Often, after working through my emails, pricing up additional works that the client wanted doing, I might decide to take a stroll across the camp, to see how we were doing on the various buildings that we were working on.

It would take a couple of hours just to walk around; the place was that big. I'd drag myself from one building to the next. We might be doing a bit of roof-tiling on one, then a few hundred yards away, we might have a team pouring asphalt, or putting up fascias and soffits and guttering. On another building, we might be installing roof lights. We were doing lots of different things on lots of different buildings, some of them new, and some of them old.

There was *never* a single time when I left my office and wandered around the site that I wasn't glad that I had done so. There was always some benefit, something I'd have missed if I hadn't taken that walk. In other words, there was always a reason for having made the trip. I never knew what it was going to be at the time that I set out, but I was always glad that I had done it by the time I got back to my desk.

We probably had thirty or forty operatives of our own on site at the height of the programme. Still, with a few months to go, the Contract Manager was going to move onto another scheme to get that one up and running. The boss asked me if I thought I could handle this one myself. I said, to his delight (because it saved him a wage) that I'd be fine on my own. For the next three or four months, I was both Project QS and Contract Manager. It meant putting the lads to work, ordering the materials, and providing the RAMS (Risk Assessment & Method Statements) for the work, as well as performing my own QS duties.

As far as doing the RAMS was concerned, I wasn't really qualified to give Health and Safety advice. I didn't have the proper certification; however, all I had to do was take an earlier set of RAMS – as prepared by our qualified Contracts Manager – and adapt it to the scheme that we were about to start. It's all pretty generic. Who is providing the access, e.g. scaffold if required to the work area? How many men will be carrying out the work? How long will the job take (e.g. days or weeks), and anything else that needs to be taken into consideration to allow the tradesmen (or women) to carry out the work and head home safely at the end of the day.

The phrase 'a qualified workforce' pops up often. You basically have to cover the arse of the main contractor or client by assuring them that you've thought about the job that you're about to start, and that you know what you're doing in order to carry out the task in a safe manner. You need to have considered all of the risks involved, such as working at height.

As mentioned, RAMS is an acronym for Risk Assessment and Method Statement. You've thought about the risks, and you've considered the methods that you're going to employ while carrying out the work in order to mitigate that risk.

Every week, the main contractor, our client, called us into a meeting where we were put on the spot. We had to say what work we were carrying out on which building in the coming week. They would ask how many men were needed to undertake the planned works. They would then ask how many men we had on site: how many tilers, how many flat roofers, how many cladders, etc. The system that they had in place immediately showed up any shortfall in our labour provision. It was quite a robust and revealing system. There was no hiding place.

Then again, they were probably facing massive liquidated damages if the project came in late; therefore, it was important that they kept each of their subcontractors on programme. Every single subbie, of which we were just one of maybe a hundred, was subjected to the same weekly grilling. It was a well-run, well-managed, and well-motivated project.

It was exciting, and certainly one that would pass the 'I did that' test whenever you drove past it. And, guess what, a couple of months before the project was due to complete, two of the supposed end-users, the navy and the RAF, declined to take possession of the site.

Envisaged as the new home for Britain's three main military outfits, two of them decided that they weren't interested in using it after all! The Navy said it's miles from the sea. Why would we go there? The RAF said we've relocated to another base where we're quite happy, so why would we go there? So it is now destined to be the headquarters of the British Army and nothing else.

That all of this effort could have gone into providing a base for all three disciplines – with accommodation and support services for all three – only to leave the army sort of rattling around in what should have been a shared space, just makes the mind boggle. You put your heart into building something, then two of the three people that you built it for don't even want it!

As a taxpayer, I should be shocked. As a construction person, I'm not surprised. There's many a slip between cup and lip. Construction is an arena like any other. Sometimes, you just have to accept that it's a mad, mad, mad, mad, mad, mad, mad world. But at least try to be good at what you do yourself. It's all that we can ask.

CASE STUDY TWO. LANARKSHIRE HOSPITAL

There are many ways to fund a project. Imagine you want a new kitchen. Can you afford it? If you can, you take a trip to the local showroom, choose a kitchen, find someone to install it, and then you take out some money from the bank and hand it over. You've just funded your new kitchen. If you haven't got the cash, you can always take out a loan or maybe pay for it in instalments.

Now imagine a local authority needs a new school, or a new hospital. All of a sudden, you're not talking about a few thousand pounds, you're talking millions, or hundreds of millions. In the early part of the 21st Century, a couple of new buzzwords started doing the rounds. They were acronyms and abbreviations, such as PPP and PFI.

PPP is public/private partnership. This was where government money (i.e. our taxes) was married up to private money (e.g. rich investors) to fund certain expensive projects. Say both parties put up fifty percent of the cash. The government would take their half out of the tax pot, and the investors would put up fifty percent in return for interest on the loan and repayment over several years. It meant the local authority only needed to find half of the money at the outset of the project. The rest could be staggered over a number of years.

Imagine if that local authority had none of the cash, yet still needed that new school or new hospital. This is where the PFI venture came in. All of the funds would be raised privately, and the local authority would pay that money back over decades, not just a couple of years.

One such PFI scheme was a £300 million hospital for the Lanarkshire region of Scotland. When I was working for the Claims Consultants – the taxi rank for QS's – I was asked to go and price some late design changes to the scheme.

Basically, £300 million was a roundabout figure. Before the contractor and the funders would put pen to paper and sign the contract with the local authority, they wanted to know if the project was £300 million, or maybe £305 million, or even £310 million. We're not talking small potatoes here. That's an extra £10 million in the back pocket of the funders and the contractors building the thing.

What happened was, the client (the local authority), as the project was about to go live, started to really analyse the nuts and bolts of what it was that it was buying. Have we got enough toilets on this ward? Do we need a bigger canteen or kitchen? They started to request small last-minute changes. These would make the difference between the £300 million and the £310 million figures.

But the contractor, Laing O'Rourke, was already flat out trying to get their heads around the project and get the deal across the line. They had no spare capacity to price these late changes. They called up the cab company, and asked them to send an emergency QS. That was where I came in.

Did I fancy a month or two in Scotland? Well, I didn't mind at all. I got the train to Stirling and found a city centre hotel, plonked my bags down, and went off to meet the client. At first, they were a little suspicious. They had a tight-knit team. They were really close to signing the contract. The last thing they wanted was a bull in a china shop upsetting everything.

They wanted someone who would go about their work with a minimum of fuss, and who could self-start. They had no time for hand-holding. There would be minimum supervision, but maximum scrutiny of whatever I came up with. After all, the client wasn't just going to hand over an extra five or ten million to the contract sum without seeing some evidence for the costings.

I was given a desk from which to work, and handed a document outlining forty late design changes. All I had to do was to put figures against each of them and justify whatever costs I came up with.

Say, for example, a storeroom was going to be changed into a WC. Well, there would be a little add and omit exercise to do. So, it's no longer going to be a store-room, so we don't need the cost of those shelves. Omit £500. But it's now going to be a toilet, so what do we need? Take a moment to think about it.

There. Done it? My list would be something like this.

Toilet and wash-hand basin. They're the obvious ones. But a toilet is never just that. You need a cistern, you need a seat, then there's little things like the toilet-roll holder. How about a towel rail? And what's on the walls? In a storeroom, it might be bare walls, or barely-painted walls. In a WC, you need ceramic tiles, certainly as a splash-back, but maybe elsewhere as well. And maybe some washable vinyl paint. And what's on the floor? In a storeroom, you might leave the concrete slab as it is. In a WC, you want non-slip vinyl.

You omit £500, but you add back a whole lot more. Maybe £5,000 more. That's your late design change costed. And there are forty more items like that one to price.

When the client asks how you got to the £5,000 figure, you show them the breakdown. A loo, a cistern, etc. All of the items listed above. They might not have considered them. They might have forgotten that you need the towel rail, or the ceramic tiles. But it's all cost. It's all claimable. If you don't charge for it, and it's needed, then the cost of it is coming out of your own profit.

Better to get these things cleared up before you set off on the journey to build your project. That way, you can sign the contract at the correct sum, and there's no confusion and everyone's happy.

I must say, in all of my years in construction, I've never come across a better project team than this lot. They were all hard-working, hugely capable, and highly motivated. It was a privilege to be amongst them. Once I had completed the costing exercise, they shovelled a few other duties on to me. They asked me to work out a cash-flow exercise for the whole project, the £300+ million. They invited me into client-contractor meetings. They made me feel a part of it. And I was only there for six weeks. It was great. But my work was soon done, and then I headed off to my next assignment.

Recently, following the collapse of major contractor Carillion, the new Royal Hospital in my home city was mothballed just months from completion. Laing O'Rourke have been brought in to finish the project.

I know we are in good hands.

PRIVATE DEVELOPERS

It took me twenty years of Quantity Surveying before I finally came across that rare beast, the Private Developer. When I finally did so, I have to say that I have very mixed feelings about them. In fact, two of them will appear in these Case Study chapters, so memorable (and for all the wrong reasons) those experiences were.

Now, by far the major employers for a QS are the contractors: be they small, medium, or large. They probably employ eighty percent of Quantity Surveyors at any one time. Then there are the private QS's. These maybe make up ten percent. That leaves five percent freelancing, and five percent working for private developers. These figures are approximate, but they should give you a feel for the opportunities that exist for you.

For two decades, I blissfully went about my business, unaware that these private development people existed. These are the investors, the speculators, and sometimes the downright dodgy people that are drawn to the world of property development. They see it as a good way to make money, and there's nothing wrong with that, but there is something of the snake-oil salesman about a lot of them. They (and I'm just talking about the bad ones here) are the people who roll into town, pitch up a tent, and draw you in with their get-rich quick schemes. They are there to spend other people's money to finance their own best-deal-ever scenarios. Just don't expect to find their tents there in the morning, because they'll often take your cash and run.

In fact, I'd even go as far as to say that some of them are gangsters and criminals who are looking to go legit. Nothing wrong with that. They reach a time in life where they're now married, have a family, and no longer want to live in fear of that five a.m. knock at the door when the long arm of the law finally catches up with them. So they take their ill-gotten cash and put it into something legitimate. They become property developers, a term so broad that it could mean any number of things. It's a good 'hide-behind' and allows them to talk about what they do, should they ever rock up at a dinner party with a bunch of straight people.

Now I realise that I'm doing a massive disservice to the many good people who are entrepreneurial and have a dream – a vision – to build something where previously there was nothing. Maybe they buy a small bungalow on a large plot of land. In their mind's eye, they can see the potential in the plot. What the area is calling out for is a block of flats, maybe a retirement home. As it currently stands, the plot is wasted being home to just one tiny bungalow. It's time to unleash that potential.

They buy the plot, apply for planning permission and, once granted, find a builder to construct whatever it is they're planning to build. This gives work to architects, to trades-people, to site managers. They buy materials from local merchants. The world turns. Everyone earns. Nothing wrong with any of that. These are the legitimate schemes that rightly take place amongst the myriad construction schemes that make up our daily lives. But beware of the charlatans. Beware of the slippery, fork-tongued men and women who sell you their dream at the cost of other people's. You'll be tarred with the same brush for having believed them. Remember, if something sounds too good to be true, it usually is.

Keep your wits about you. Tie yourself to the mast of your Hippocratic Oath and don't fall for their siren-song. You're a QS. You're your own QS. You're not there to lie and steal and cheat on behalf of some 'fancy Dan' in a fancy car living in a fancy house, usually funded on the misery of others.

How many years has their fancy-named company been trading? Eighteen months? Smell a rat. Not published any company accounts yet? Smell a rat. They're thirty, forty, or fifty years old, but they've only just started the business? Ask why. What were they doing previously? What was their company called previously? They probably won't tell you. Do some digging. Did they go bust owing millions, only to start up the next day with a different name? Sometimes, the new company name, complete with a new website, miraculously rises from the flames within hours of the previous company's demise. How does that happen? Because it was all part of the plan. Or if not the plan, then certainly the contingency plan.

That million pounds that they owed when they went into administration, look down the list of creditors and you could weep. There, amongst the victims, will be that local builder's merchant, a small-time plumber, the bricklayers and plasterers who came in good faith and were each left being owed thousands of pounds. You have twenty suppliers and small contractors owed fifty grand each. That adds up to the million they owed when they went bust. That money sinks companies. It loses people their houses. It loses people their jobs. It costs marriages, and destroys families.

But guess who doesn't lose their house, or their car (it will all be in someone else's name)? Yes, the property developer. The bad property developer, I should say. I excuse all of the legitimate ones, of whom there are many, and apologies to all the good ones out there for the nature of my rhetoric.

So, as a QS, if you do go down the route of working for a private developer – and there's no real reason not to – just keep your wits about you. Look over your shoulder or, even better, have eyes at the back of your head, as well as in front of you. Remember you're a QS. You're a construction industry professional. Do the right thing. And don't buy the snake-oil.

(And if you're looking for examples of what I'm talking about, continue to read the case studies further on in this book.)

LIFESTYLE DEVELOPERS

A small contractor recently introduced me to the expression 'Lifestyle Developers' to describe the kind of housing developer that gives the market a bad name. Essentially, it means someone who uses the role of developer – building things for other people – to fund their own lavish lifestyle that neither their talents nor work-ethic merit.

One such scheme concerned the building of a fabulous 'White House' style mansion for a very famous footballer. I was QS for the roofing company putting the roof on this ostentatious palace.

Now the client was a former England footballer. He was, in fact, a bit of a hero of mine. But he was being taken for a fool by this Lifestyle Developer. This famous footballer had certainly chosen the wrong guy for the job.

I can imagine, when he decided to build a new multi-million pound home for his family, that he asked around for the names of some development companies who could manage the project. You'd want someone with a track record and a good reputation. He might have asked some of his footballer friends for the names of the people who'd helped build their own luxury homes.

He probably ended up with four or five names. He probably called them up and asked for a meeting. That meeting would essentially have been their interview for the role. Then, the footballer probably went away and thought about it.

The price would probably have been the same from each of the candidates, or there or thereabouts. Say, the build cost, plus a fifteen or twenty percent mark-up. The deciding factor, in choosing which property developer to entrust with the building of the client's new home, would not therefore have been just about price. It would have been trust. Maybe trust and likeability. Who did he most trust to deliver the project on time and on budget? Who did he want to work with, because the building was probably going to take a year to complete. Who did he want to spend that year with?

From the four or five names he initially came up with, he probably narrowed that down to a shortlist of two. He surely spoke to his wife about it. They would have decided which one they most liked and which one they most trusted. Then they made the call and told one lucky punter that he'd got the job. Not a bad one for the CV, for someone. I built so-and-so's house.

Except they chose the wrong man. They fell for the spiel. They bought the snake-oil. When the chap turned up in a fancy car for their meeting, the client probably thought that the guy must be good at what he did because he was showing all the signs of success. Well-dressed, nice watch and phone, nice car. But it was all show. The guy was a con-man, or at least a bull-shitter. He was, to all intents and purposes, a fraud. Still, he looked the part, talked the talk, and got the job.

Say the build-cost was £2.4 million. Say the project was slated to last twelve months. Logic tells us that the developer should have been drawing down two hundred grand a month from the client. In twelve months' time, the house would be built, the contractors would all have been paid, and the developer would have made a tidy profit. Everyone's happy. Everyone's a winner. But that's not what happened.

Because the developer is living beyond his means, he's living beyond his earnings. He probably had his own fantasies about being a Premiership footballer and living the associated life. But he wasn't a Premiership footballer. He was a builder. He should have known the difference but he didn't. Ideas above his station.

So the client would hand over his two hundred grand every month, and the developer probably kept a hundred grand of it for himself. That only left one hundred grand to get the work done on site. And it wasn't enough.

At first, an army of small contractors was drawn to the project. They started laying foundations. They started building the walls. They put the roof trusses on. We, as the roofer, begin laying slates. At first, the developer has a little grace. We have to do a month's work on site before we put our first application for payment in. He's then got another month in which to pay it. But, after two months on site, when all of our invoices start to go into arrears, or it becomes apparent that we're not being paid – when the developer stops answering his fancy phone – the work on site slowly grinds to a halt.

Why are there no bricklayers on site? Where are the plasterers? Where are the roofers? The answer is that they're all working on other jobs where they either are being paid or at least they still have that expectation. Because, guess what, they all have families to feed and mortgages to pay. They need to be paid, and if they're not, they go elsewhere. Not hard to get your head around, is it?

Of course, the client is blissfully unaware of all this. He's still handing over two hundred grand a month. After all, the developer is still taking *his* calls. He's still smiling and having a laugh with him. They're like best buddies for the duration of the project. But it's all a sham.

And, as actually happened on this scheme, when the footballer's wife says she's coming to the site on a particular day, to see how her beautiful new house is coming along, what does the developer do to disguise the lack of activity on site, the fact that there's no one actually here doing any work? Well, he employs a couple of agency labourers for the day and gets them to move left-over materials from one spot to another in a completely pointless exercise, all designed to impress the lovely lady and show her that some work is going on. Yes, it's really important that these pieces of timber are being carried aimlessly from point A to point B. That's progress right there.

She's probably seen the building change a little since she was last here. She probably doesn't know a great deal about construction so she won't be able to ask the pertinent questions or sniff out the fact that it's all a masquerade.

She goes away, fairly happy. She can, maybe, see that they might not be in the house by Christmas, as she had hoped. The project is running a bit (actually quite a bit) behind. The developer blames the shocking bad weather that they had a couple of months ago. It's caused the delays that even she can now see on site. But, he assures her, she shouldn't worry too much because he's on the case. It will all be fine in the end. And can he have his next two hundred grand, please.

And the project limps along. The developer may have an army of small contractors on his back chasing payment. Those tradespeople and subcontractors probably won't be coming back to the site until payment has been made. But there's a cladding contractor that he hasn't ripped off yet. He can get him to site to carry out some work before he himself goes the way of all the others.

So the house starts to get constructed out of sync. For ourselves, the roofing contractor, we withdrew from the site before we had finished all of the lead flashings to the parapet walls. Quite important, those flashings. They basically plug up all the gaps. But we hadn't been paid for the work we had done, and we certainly weren't going back to do any more work and leave ourselves even further in debt by forking out for yet more labour and materials. So the gaps remained exposed.

What does the developer do? He goes ahead and puts the capping stones on top of the parapets. When the client or his wife next comes to the site, it will look like progress. But it's all a sham. The roof doesn't work. When it rains, all of the water will drive inside the house, because the gap between the slates and the stones has not been plugged.

Eventually, the project limps across the finishing line. The footballer and his family move into their fabulous-looking new home. And then it rains, and they all get wet. And the client (the footballer) phones us up and complains that his roof is leaking.

'We know,' we respond.

Someone from his organisation phones me up and asks if we can come back and fix the roof.

'Only if you pay us,' I say.

'But we've already paid for the roof,' they reply.

'No, you've paid the developer. You haven't paid us. And neither has he. We haven't been paid. You need to get on to *him* and get *him* to fix the roof. Oh, and one more thing, the next time you're getting some work done, if the builder turns up in a better car than your boss, worry!'

Lifestyle developers. Developers who are merely in it to fund a lifestyle that they can neither afford and which they don't deserve. Watch you don't get to work for one, because you will be a party to the ripping off of honest, hard-working tradespeople and suppliers. You'll forever be ducking and diving. You'll stop answering your phone. And you'll no longer be a construction professional. You may as well pitch up your tent and start practicing your spiel. Time to start flogging the snake-oil.

CASE STUDY THREE. TWO RESIDENTIAL APARTMENT BLOCKS

I like to think I've got quite a varied CV. Like I say, I've been the guy asking for the money, and I've been the guy paying the money. I've worked for small, medium, and large contractors. I've worked in private practice. Even done a bit of freelancing (several years, in fact). One thing that wasn't on my CV, however, was a new-build residential scheme. For a, yes you've guessed it, property developer.

There's always a slight apprehension when you take on something that you have no direct experience of. That's only natural. You doubt yourself a little, and wonder if you've got what it takes. But I'd been in that many different QS arenas that I thought I'd do just fine. Same skills, just a different setting.

Basically, this developer (actually a group of three like-minded individuals) had identified a piece of land for sale. It was in a residential area, and had formerly been home to a factory. The site was a corner plot, rectangular shaped, about 80 metres long by 20 metres deep. It backed onto a railway embankment. The train station was less than a minute's walk away. There was a historic park nearby. It certainly had its selling points.

It seems the trick with developing is to sell your idea to a bank or a private investment company. Borrow money at a reasonable rate, and build the thing you want to build using someone else's money. Then, you sell the thing, give the investors their money back, and the difference between the two sums is your own profit margin. All you have to do is find someone with money to invest and convince them of the value of your idea and your ability (or even better, your track record) to see it through to fruition.

In this case, this developer (newly established, no accounts yet published) had managed to secure £2 million in investment to build 47 apartments in a two-block new-build residential scheme. After all, there was a train station nearby, a historic park over the road, and it was only ten minutes by train into a bustling city centre. There'd be no parking issues. A really easy commute to work.

Would I come on board and place the various sub-contract packages, such as groundworks, bricklaying, flooring, mechanical, electrical, roofing, joinery, plastering, screeding, painting, and decorating? Why not, I thought. After all, this was my year of getting sacked. I needed a job.

I'd be working alongside a site, stroke, project manager. He would be permanently based on site, five days a week, from 7.30 a.m. to 4.30 p.m. Building sites tend to start and finish a little earlier than your regular 9 to 5 office job. No bad thing if you can get out of bed. Finish early. More of the evening to enjoy.

Most sites are secured by a surrounding of hoardings, and within the compound, there is usually a site office, a storage unit, a canteen (of sorts) and a toilet block. Ours was no different. I would spend two days a week on site. The scheme had a budget (£2 million) and I had to spend it as wisely and conservatively as I could. I would do this in conjunction with the site/project manager.

Now you know that I have my reservations about developers. These blokes were nice enough, but I'm a QS first and foremost. All I could do was my job as well as I could. I was well to be wary, because this scheme had already had its fair share of setbacks.

Initially, the developers had brought in a small to medium-sized contractor to manage and build the development. They'd have experts on board to build it. These contractors would probably bring in a lot of their own trades-people and subcontractors. It would have been an efficient way to construct our two residential blocks. Except the developer and the contractor fell out. It may have had something to do with money. I don't know, it was before my time on the scheme, but it seems a good bet. Maybe the contractor wasn't being paid on time, or maybe the developer wasn't happy with their performance or progress. Anyway, there was a parting of the ways, and the project was mothballed while the developers sought to re-finance the scheme.

Eventually, after almost a year in hibernation, a new set of funding was in place and the scheme was revived. This was where the project manager and myself came in. The developer was essentially going to employ their own people to do the job of the contractor. The site manager (full-time) and me, to save money, on a part-time basis.

The timber frames for the two apartment blocks were pre-fabricated. The order had been placed months earlier, due to the long lead-in time. It may even have been a leftover from the scheme's first incarnation. Similarly, the groundworks contractor was 'grandfathered' to the scheme, although there were variations and some re-mobilisation costs to be agreed.

What did we need now? We urgently needed a brickie. Did I know anyone? Did the site manager? He lived locally to the project. He also knew a brickie. We brought the guy in. He was great. He agreed to come on board. First subcontractor ticked off the list. Now, who's next?

We started to look at the various work-packages. We'd try and come up with three or four names for each. We might both throw two names into the hat. We'd give them a call. Invite them to site. Have a chat. Give them a set of the architect's drawings. Give them a specification and a Bill of Quantities if we had one. If not, I'd do my best to quantify what we were asking them to do.

Now here's the thing. Just because you ask someone to price some work for you, it doesn't mean that they will. They'll usually ask who the client is. As soon as you say that it's a developer, half of them immediately hang up the phone or they make some excuse about being too busy at the moment to take on any more work. The last part might well be true. They are working for people that they know and trust, that they know are going to pay them, which then allows them to pay their own staff. Why risk it to go out on a limb to help someone else who might fulfil their vision at your expense?

I was well aware of these subcontractors' reticence, or quickly picked up the vibe, and I did my best to assuage their fears. I'd basically say something like this.

'Look, I don't know these people from Adam, and I know how you feel about developers. All I can say is that I am here as much for you as I am for them. This work needs doing. There is a job here for someone. If you take it on, I will always answer the phone to you, and I will fight tooth and nail to make sure that you get paid. On that basis, are you interested in tendering for the work?'

More often than not, those wary contractors – as skittish as a deer in the woods at the start of the conversation – could tell that I meant what I said and they would give us a quote for the job.

But still, we had problems. No wonder the previous contractor was having trouble getting paid (if indeed that's what it was), these developers were like kids in a candy store. Every time they spotted another potential opportunity, they would dip into the pot of funding for this scheme and buy up another development. All of a sudden, our scheme had a £300,000 shortfall because they'd bought a big massive house ripe for refurbishment and conversion to apartments. When the funders came to the site to see what progress was being made, someone had to explain why we were behind. But don't mention the three hundred grand that's gone missing from the budget. Time to drag out that bad weather excuse again.

Best of all, there was a problem with the setting out. I knew something wasn't right when the engineers came out and moved their setting-out markers three times. I didn't know what was wrong, but I knew that something wasn't right. But what did I know? I was only there part-time, and my role was to place the orders with subcontractors. That was hard enough, given how uneasy they were, but we seemed to be making some progress on that score.

Still, whenever I thought about the scheme as a whole, I was convinced that something wasn't right. The brickwork between each apartment was different. I didn't know much about brickwork, but why have four different scenarios for what was essentially the same party wall between the separate apartments?!? The brickies were a decent bunch. They were grafting, but their instructions would change on an almost daily basis. They didn't know what was going on either. They'd build the walls in the foundations so high, only to be told to take them down again. My confidence was draining in the project manager. He was on site five days a week. His window in our office overlooked the site. He could see everything that was going on. I was there two days a week, facing the opposite wall, placing orders for sub-contract packages. I wasn't in the dark, but I did not have the handle on the project that this man was meant to have.

I was about to go on holiday for a week. I thought about calling one of the developers. We were due to have the prefabricated timber frame delivered during the time that I would be away. Given the different scenarios with the brickwork, and the fact that the site engineer had moved the goalposts three times, I had real doubts that the frame and the footings would marry up. I wanted to voice my concerns. Basically, I was convinced that the site manager had somehow messed up. What I wanted to say to the developer was 'Look, I'll be very surprised if, when I get back from holiday, this thing has all come together. I don't know how. I don't know why, I'll just be very surprised if this scheme is still on track by the time I come back.'

Yet I did not make the call. Why? Because it obviously meant that I would no longer be able to continue on the project. It would have been a statement that I had no faith in the project manager, and that I could no longer work with the guy. One of us would have had to go. And, to be honest, it would have had to be me. He lived around the corner from the site. He was there at 7.30 every morning. I would not have been able to do that. I wasn't qualified to manage a site. It meant I would have been the one to go, and I didn't want to just walk away from a job. And, besides, I could have been wrong. It could all go swimmingly. I might return from holiday to find the frame in situ, the project flying, and my confidence in our site manager renewed.

I went on holiday. I came back. I went into work. The timber frame was up on one of the buildings. I climbed the stairs to our office. I sat down at my desk. And I casually asked how things were going.

'Okay,' came the reply.

Then, even more casually than my own question came the following comment.

'The building's in the wrong place.'

'What!'

'Yes, one of the buildings is in the wrong place……'

I would have liked to have seen the look on my face. I was, I think, speechless.

Eventually, I managed to say 'How come?'

'The architect's fucked up.'

Great. We've got someone to blame. Not 'we fucked up' but 'they fucked up.' And they had, to some extent. The coordinates they gave on their drawings put the second plot too close to the railway embankment. Basically, we would step onto land that belonged to the good old British railways. There are strict rules in place about how close you can get to train-lines, because trains vibrate, and buildings don't like that sort of thing.

So the architects had put some wrong numbers down on their drawing, but what about us? What about the engineer, and what about the groundworker, and why had no one spotted that the troublesome building was set back about a metre further from the road than the other block? Apparently, they were supposed to be in line. Our site manager had a lovely window *in line* with the foundations of both blocks. He looked out of that window fifty times a day. It's not good enough to just say that it was all the architect's fault. We can all make mistakes. That's why we have teams. Hopefully, someone can come to your aid.

Do you know how difficult it is to deal with the railway companies? They have fifty different stakeholders. They meet about twice a year because it's so difficult to get them all in one room. And when they do meet, they've got a lot more to talk about than the little problem you've caused yourself by encroaching onto their land.

It was an absolute disaster and an absolute joke. My worst fears had been realised. What to do now?

Did I feel at all responsible? Of course I did. I was there. If I had maybe invested more time in the scheme as a whole, analysed the drawings, and thought about the setting out, I might have spotted the mistake. Except that wasn't what I was there to do. I had no prior experience of new-build housing or residential schemes. What I knew about setting out, which is a site engineer's job, you could write on the back of a postage stamp. And it wasn't my job. I was there two days a week for a total of eight weeks, a mere sixteen days. In that time, I placed more than a million pounds worth of subcontract orders. Yet, of course, I still felt responsible. Why didn't I take more of an interest? If you're part of a team, you win and lose together. At least that, I think, is how it should be.

I felt bad. But it wasn't my job to get that particular aspect of the building right. And if it had been, and it had gone to pot, I'd like to think that I wouldn't have been so quick to blame just the architect. Yes, they had put some wrong coordinates down on a drawing drawn many miles – and many months – away from the actual build. What's wrong with capturing that mistake out on site, especially if you're there every day? And if the site engineers are confused, and the brickies are confused, and you are in charge of that whole process, what's wrong with calling a meeting or doing some delving and getting to the bottom of what's going on. Don't just let disaster happen in the knowledge that you have a patsy lined up who can shoulder the blame. We're all in it together. Take control. Stop the mistake from happening by being the best you can be.

To compound matters, we had taken on a cleaning company; just someone to provide the paper towels and loo roll in the toilet, clean out the cabins once a week, that sort of thing. It was a mate of the site manager's. He came up the stairs into our office and asked how the project was going.

'The building is in the wrong place,' the site manager said, casually.

'How come every job we've been on is fucked up,' the cleaner replied.

Well, excuse me, but I've never really heard of the cleaner being able to sink a project. I've never heard of a scheme failing because we had the wrong colour paper towels in the bogs. But a site manager can sink a project. The man at the top, making the decisions, steering the ship.

My disappointment and lack of confidence in the man must have been written all over my face. I was embarrassed. And then I was let go. They no longer needed a part-time QS, the site manager told me. I asked who would look after the subcontract packages that I had procured. I had given assurances to these people that I would defend their interests at every turn if they came on board. Who would look after those interests now? The guy told me that they would take on a consultant instead (which is a part-time QS!). I got the message, got my coat, and went to look for another job.

Six months later, I drove past the building. It was (and still is) a ghost town, an unfinished carbuncle. One building stands in darkness, complete in all but name. Its sister building is still just concrete footings in the ground. Pallets of bricks stand nearby. A footprint in the ground, too close to the railway tracks, another failed project. Oh, and the developer? They went bust.

CASE STUDY FOUR. NINE NEW-BUILD HIGH-END HOUSES

You might think that my first experience of new-build housing might have put me off, but no. I was soon back for more. I found another two-day a week gig for another developer. I was still writing my two books, so I was still happy working part-time. Besides, as I've said, the money from freelancing is pretty good, so I could still afford to pay the rent and bills.

This time, the scheme was to construct nine luxury new-build houses in a cul-de-sac location in a nice commuter village. My role, as before, was to let, by which I mean award, the sub-contract packages, finding maybe two or three or four small companies to bid for each portion of the work. Again, I would pick the brains of the site and project manager (two different blokes this time) and use my own list of contacts to invite subcontractors to provide us with a tender. It was going to be another difficult sell, because once again the project was beset by funding issues.

Basically, there were three of us in the site office, the project manager, the site manager (both full-time), and me doing two days a week. Plus we had a telehandler driver (like a JCB) / labourer to move stuff around the site and to wait on the tradesmen and keep the site tidy, etc. Except there were no tradesmen on site. Because those who had been there previously had not been paid and therefore they weren't coming back until they had been.

There were more of us sat in the site cabin than there were out on site doing the actual work. And remember, those cabins you're sitting in are all part of your prelims cost. And the site management team sitting in them are all part of your prelims costs. And whereas these costs might usually be ten percent of your overall budget, if your scheme is going to overrun by twelve months, and you're spending ten grand a month on the cabins and the people sat there, wistfully inside them, that's £120,000 wasted because you didn't get your funding in place or didn't pay it out in timely fashion. It's an absolute waste. It makes no sense, and smacks of an ostrich-like head-in-the-sand mentality by the people who are ultimately in charge… the developers.

We needed the scaffold to be altered in order to allow the next phase of the brickwork to progress.

'When are you coming back to site?' the project manager would ask the scaffolder.

'When I get paid,' would come the response.

What can you say to that? There's no argument that you can level at the scaffolder that is going to make him come back to the site and carry out more work when he hasn't been paid for the last lot. Is there? So the site manager and project manager would bemoan the lack of progress. Their hands were tied. Meanwhile, I was actively trying to round up more lambs for the slaughter. Often, when I'd ask 'Does anyone know any electrical companies, or roofers, or plumbers?' the answer I would get is 'Yes, I know someone, but I'm not bringing them on here.'

I adopted the stance I'd had previously. I knew companies were skittish. I promised to fight their corner. I offered them payment terms that wouldn't expose them to too much risk. I knew there was a real chance that these guys would not get paid. The best I could hope for was that they might get paid eventually.

Ordinarily, even in a best-case scenario, subcontractors would do a month on site and then submit their first invoice, and then wait a further thirty days to get paid. That's two months before seeing a penny for their outlay. And that's if you get paid on time. And lots of companies have thirty-five days, or forty-five, or sixty-day payment terms from the date of invoice. That's three months before you're even *due* to get paid.

What I proposed to each subcontractor was that they come on board and submit their first invoice after fourteen days, and that the first invoice had to be paid within the next fourteen days. That way, they would only be exposed to one month's outlay. They would effectively be paid one month in arrears, with payments on a monthly basis thereafter.

That seemed fair to me. It allayed these people's fears. And the last thing that I wanted was for them to get ripped off. It also meant I could sell what was, undoubtedly, a very hard proposition. I even received an email to that effect from one subcontractor that I approached. They declined to offer a price for the work because of our company's poor credit record and our dismal reputation locally with both suppliers (of the materials we were buying) and subcontractors. What could I say? I knew it was pretty much the truth.

Those contractors that I could entice to site were taken aback by the fact that there was no-one around. This was a building site, about three-quarters away from completion, yet there was no one on site. It was a ghost town. I had to use all of my powers of persuasion to stop them from turning on their heels as soon as they walked through the gate into the compound.

Remember, these people aren't stupid, and they've been around the proverbial block. They're running small businesses, employing handy lads who want paying every week. It's part of their job to sniff out trouble and to head it off at the pass.

Our brickwork contractor was great mates with the site manager. That was the only reason he was still here. Well, not here exactly, but waiting in the wings, ready to return, as soon as he'd been paid what he was already owed. He said the same thing to me.

'If it wasn't for him (his mate), I wouldn't even have got as far as the site cabins,' he said.

If there's work to be done, and it's not getting done, and it was quite a nice project, then why is no-one here? It doesn't take a genius to work out that there's usually only one underlying reason. People aren't being paid. Sometimes you get a difficult site manager, or an idiot project manager, or a bothersome client, and even a three-decimal places QS, but folk will usually put up with all manner of ills as long as they can take a wage home at the end of the week. If they can't, they're off. Simple as that.

We needed a screeding contractor. The site manager, knew a good one. I knew a good one. It turned out to be the same firm, so the choice was a no-brainer. We would use them. I called them up. Yes, they'd be interested in coming on board, but they'd also heard about our developer clients, and they wanted ALL of their money to be paid in advance. They'd obviously picked up the vibe of the project, or done some background checks on the developer. Obviously, I couldn't agree to that. I negotiated that they get 50% in advance and the balance within seven days of completion. This was to be on a plot by plot basis. We only had to stump up a couple of grand at a time. The arrangement worked for both of us. They were a good outfit with a very good reputation. The payment terms where the best that I could get them to agree to, otherwise they weren't coming.

The developer I was working for had a commercial manager on their books. He was essentially my line manager. He wasn't too happy with the terms that I was offering these subcontractors. The 50% in advance was a one-off. The rest were on fourteen days to first invoice and fourteen-day payment thereafter, essentially getting paid a month in arrears. I explained that I didn't want to see anyone going unpaid. I had my own reputation and my own contacts to think about. And, I was a QS. I'd taken that unspoken Hippocratic Oath. His response?

'I just don't want you getting ripped off by robbing bastards!'

I bit my tongue.

These weren't robbing bastards. These were good, honest subcontractors. The best we could find. At least, they were the best that we could find to take a chance on our tarnished project. Some of these people were friends and former colleagues. Neither the project manager, the site manager, nor myself wanted to let them down. We wanted these houses built. It was a great little scheme. Families were going to enjoy living in these houses one day. In fact, they would be the envy of the surrounding neighbourhood. It was a little cracker. Was it too much to ask that the people actually building them got paid? I didn't think so.

We limped along. We'd get a bit of cash. We'd pay a few people. We'd make a little progress. Then the money would stop. And the subbies would disappear again. And the site team would outnumber the men on site. And the prelim costs would keep rising.

I met the developers at their nice offices many miles away. I told them that they had an air-bubble in their funding pipeline. They smiled the smile of successful people and ignored my comments. They concerned themselves instead with choosing colours for the bathrooms and deciding which kitchen range they should have, not that they could afford any of it, or planned to actually pay for it. Ostrich-ville again.

Finally, an issue came up on site. I needed a quick answer. I called one of the directors of the company. After (badly) answering my query, she asked me how it was all going. This was after about four weeks of us outnumbering the men on site by a ratio of 4:0. Not very good, I replied. In fact, more like fucking awful. Who works like this? Who expects a happy outcome from a scenario such as this? They'd SOLD the houses. People were expecting to move into them. But they weren't getting built because no one was getting paid.

Well, the commercial manager wasn't too happy with me telling the director/developer that the project wasn't going swimmingly (despite the fact that they had all visited the site, and knew as much themselves). At the end of that week, my three-month temporary work contract was due for renewal.

Guess what, at four o'clock on a Friday afternoon, the employment agency who had hired me (who I was working for on the developer's behalf) called me up to say that my contract would not be renewed and I was not required back on site.

Luckily, I was already doing some other QS'ing for another client. But it could have been a kick in the teeth. Let go at four o'clock on a Friday afternoon.

Anyway, I stayed in touch with the project manager and the site manager. Both left the project about four weeks later, appalled by everything that was going on and no longer wanting to be a part of it. With no more available scapegoats to blame for their rapidly sinking project, the developers soon set their sights on the commercial manager. He left in a hurry about two weeks after that.

Oh, and the developer? They went bust, owing more than £2 million. I got hold of a list of creditors. The contractor they brought in to continue the scheme got taken for more than £50K. Several employment agencies, who'd paid staff on the developer's behalf, got taken for about the same. There were suppliers, subcontractors, manufacturers, myriad companies who'd all tried to help and had paid for the privilege.

I was pleased to see that, apart from the roofing contractor, who lost about ten grand, none of the others that I had brought onboard lost out.

You learn as you go, I suppose. But the people who do this sort of thing rarely get called up on it. Instead, they continue to live in their nice houses, and drive their nice cars, and send their kids to private school, and take their nice holidays.

So, whoever you go to work for, remember your Hippocratic Oath. You're bound to need it.

CASE STUDY FIVE. A £6 MILLION NEW-BUILD HEALTH CENTRE

So, I was freelancing. I went for an interview with a medium-sized contractor. They needed a project QS. Like the name implies, this is the term given to a QS for a specific project. They offered me the job. I took it. It was slightly out of my comfort zone. Site-based, when by that point I was used to client-side, office-based QS'ing. Anyway, it was real world stuff.

The scheme was already up and running. And it was already a month behind programme. Understand that the programme is pretty much everything to a construction scheme. Budget and programme, both achieved together, is nirvana. It should be your goal, that of your clients, and the objective of everyone onboard for every scheme that you ever work on. Achieve that, and you will be beyond reproach. People might still gripe, but achieve both – believe me – and you'll have the upper hand in any argument; you will have a long and unblemished career of which you can be proud.

On this scheme, the contractor brought in a new project manager. They had obviously realised that they were heading in the wrong direction and acted quickly to stop the rot. Good on them. It was the right move.

You already know the value that I put on communication. This could have been a toxic job. A month behind schedule only two months into the scheme. The project manager, just got on with the job. The deficit was not his doing and not his concern. He just had to build a health centre.

If he could have regained the lost weeks, he would have done, but that was never the objective. He was a clever man for realising that. There was a reason that the project was behind but we'll never know what that was because we're not paid to run backwards with the ball. The best thing to do was to not be overpowered by the situation, but to do the job right from here on in. We never recovered the lost month, but we also never lost any more time. Just do your job the best you can and the truth will set you free.

This was a real-world building site. It was something that was coming virgin out of the ground. It needed temporary electricity to allow the building work to take place. That's temporary supplies. You need site-cabins. Something to operate out of. You need a canteen. Loos, for sure. Place of all these things in a corner of the site where you don't have to move them while you're in the middle of the building phase. There are contractors constantly coming and going; labourers, clients, plant-drivers. A macho symphony. That's what a building site usually is.

For myself, my role involved procurement; everything from perimeter fencing, skips to clear rubbish on site, and even a water cooler for the office. I was buying stuff. That included the remaining subcontractors we'd need to finish the job. Use your contacts. Use your charm. Use the Yellow Pages if you have to. Ask around. As long as your company doesn't have a toxic reputation for non-payment, you can buy stuff in their name. If it does deservedly have people running for the hills, then you don't want to bring them in anyway. You need to get yourself out of there pretty quick too.

But buying stuff is fun, as long as you know that your employer is going to back you up by actually paying for it. If not, it's a poisoned chalice. I always try to be fair to both buyer and seller. I think you should be too. In fact, I insist on it! To take stuff and not pay for it is just theft. Try to be as honest as you can while we get things built.

We delivered a new-build health centre, loved by the community, no later than the one-month behind from when we had inherited the scheme. Some days, the project manager would say to me, 'Let's take a walk.' He wanted nothing more than a mate to walk around the site with, so sociable was he, but I could still contribute and pick up insights on the project, as I always did. Every walk around was me doing my job and being a good QS. The fact that your job can be fun and you can see the master at work, well, that's just your privilege in the role of a QS. That's what it's all about. That's what you studied for. That's why it's a hell of a job.

THE GAME OF CHESS

You might think, as a QS, that you are joining a club – a band of brothers and sisters. More fool you, if you think so. You're essentially entering a chess competition. Other QS's will be your opponents. Like it or not, we are all there to serve our own clients and paymasters. On a bad day, I despise other QS's. They will try to tell you that the earth is flat in order to win an argument and withhold payment. At times like that, I want to weep.

These guys aren't your friends, they are your competitors, and you've agreed to take them on in a game of chess. Now the good thing is this: it's a game that you can both win. How? By making the client stump up additional cash for *legitimate* variations. If not, then you're both fighting over the same meal, and someone is likely to come out of the scrap slightly better than their opponent. That's the winner. That's the better QS.

The way it works is this. The main contractor asks for some additional work. You say it's going to cost five grand. They value it at three. The middle ground is four. You have to out-think and win-over your opponent and convince them of the validity of your position. Use precedent, tug on heart-strings, out-manoeuvre, do whatever you have to do to ensure that you come out slightly ahead. That's how you pay your own wages. That's how you earn that pay rise or bonus.

The best way to win the fight? Get the project into your DNA. Understand it, so that you're well-versed in its every nuance. Talk to the lads on site. You'll know everything that's going on.

I once had a main contractor's QS tell me that he didn't want to pay a grand for an extra bit of work. He said, and I quote, 'That was two bits of fuck all.' He acted like he was offended that I had the cheek to even charge him.

My response? You wanted some additional hand-railing putting up on a roof. Two men drove down from Manchester, a distance of 200 miles, on a Friday, to carry out the work. The materials were delivered from Kidderminster to the job in Wiltshire by two men in a flat-back truck. The work was then carried out on the Saturday, so the men from Manchester stayed overnight in a hotel. That's four men, plus materials, plus accommodation, plus delivery. That's a thousand pounds. What could he say? He knew the work had been carried out. He knew that it had been done at his request. He couldn't deny my knowledge of what had taken place. I won my game of chess and got my grand.

That same main contractor's QS, when I gave him a cost for a variation, some additional work, would often ask for a breakdown to show how I had arrived at that cost. I might say, well, I've allowed for two days labour, for two men at £150 a day per person. That's £600. I've allowed £200 for materials. I've added £200 for our mark-up. The cost is a grand. He might say, I think your labour cost is too high. Well, guess what, it might be. But it might also be too low. If we do it in a day, I make a bit more profit. If it takes me three days, because it rains or we get held up on site through no fault of our own, is he going to say, have a little extra money? Not bloody likely. So, the price is the price. It's swings and roundabouts. Remember, it's all a chess game. Try and come out on top.

Right now, I'm employed in a position where I don't get to site. I'm a Quantity Surveyor performing the role of an estimator. I price jobs, the quotes leave my desk, and at the end of every month I invoice for the work carried out. And, guess what, I'm a sitting duck. I'm a soft touch. I have very little knowledge of what's going on out on site. All the little extras get done with little or no recompense. The lads dance to the main contractor's tune on site. I ask for our money, and very little else. At best, I get what I'm given. I'm not playing very much chess these days. As a result, I think our company is failing to maximise our returns from every project. I'm no longer 'Fights-with-QS's.'

This is not to be adversarial. I still believe that everyone involved in a construction project, or in any project, come to that, should want the same thing. We want the thing to be built, and we want it on time and on budget. It's nice to be able to meet up with colleagues and associates socially, and to be friends and have a laugh where possible, but your job as a QS is primarily concerned with the costs of construction, and you would be remiss in your responsibilities if you didn't try to win more games than you lose. It's not a nasty game, it's still only chess;

but try to out-think, outsmart, and outmanoeuvre your opponent. Kill them with kindness if you have to. As long as you win. Or as long as you both win.

And remember, Right is Might.

THE FUTURE OF THE CONSTRUCTION INDUSTRY

I'm fifty-three years old as I write this, and I can clearly remember two major recessions that affected the national economy and saw construction grind to a halt.

When I started out in the building industry, times were hard. I left my home up North, and moved to London. I would literally walk from street to street, looking for the tell-tale sign of scaffolding outside a building, and walk up and ask for a job. I had lots of them. Some lasted for a week or two (if I was lucky), while I was also told 'nothing doing' on hundreds of occasions. But I kept looking, kept asking, and eventually I fell in with a roofer who was good at what he did.

By word of mouth, he seemed to have one job after the other. It kept me in employment for about five years. Then, one day, it all changed. The optimism vanished from the air. Everyone ran back to their caves, scared, unwilling to venture out. It was dog-eat-dog, hand-to-mouth. It was almost apocalyptic. This was the late eighties / early nineties. The wheels just fell off the economy, and it seemed like no-one had a penny to spend on anything. No one was prepared to venture. No one knew how to gain. The best you could hope for was to scratch around for scraps in order to survive.

It was at this point that I decided to study Quantity Surveying. Some people questioned what I was doing. 'Why are you studying for a job in the construction industry when there are no jobs?' I thought, surely this is the best time to be studying, when there are no jobs. Then, hopefully, when the economy picks up, I'll have the qualifications I need to get the job that I want. It was one of the best decisions I've ever made.

Low and behold, just as I was about to graduate, the economy started to pick up. I remember a lad I was studying with, just before he graduated, phoned around a few construction companies to ask if they had any vacancies. He got a job as an Estimator. He'd studied Building Surveying, so he wasn't exactly ideal for the role, but they needed someone, and he needed a job, and so they took him on. It was a good indicator of the state of the economy that I was about to emerge into. Hope was returning. And so it proved.

I had continuous work for the next fifteen years until, guess what, the economy crashed again. Where did it go? Where was all the money, the optimism, the jobs? It was tragic. Blame the bankers, or whoever you want, but the only thing that we can all agree on is that practically everything stopped. There were precious few jobs to be had. Personally, I took up residence in an Estimating office (after about a year of finding nothing). It was good practice, as Estimating is one of the five hats that you wear as a QS, but it left the other four-fifths of my role going rusty. I always knew that when the market picked up, I'd be off. After four years in the doldrums, the green shoots of recovery began to appear, the phone started to ring, and I eventually returned to QS'ing.

I had to blow a few cobwebs off, and it was a while before my confidence returned and I thought, 'Actually, I'm not a bad Quantity Surveyor. I can do this.' When will the next recession occur? Who knows, but it will in all likelihood happen, no matter how rosy things may look at any one time. An adage that I like, and seems obvious, although it appears to go contrary to common custom, is 'Sell when everyone's buying. Buy when everyone's selling.' That's another tip for you that I hope will make you rich.

I've always said, if I had money, I would put it into property. The only problem is, I've never had any money, so I've never been able to put my philosophy into practice. But property seems a safe bet. Unless you buy into an area that somehow turns into a ghetto, property values just go up and up. I read today that property prices had *slowed* to only 7.63 percent per annum. You can't get that in interest from the banks, as far as I know. Bricks and mortar are a pretty good bet. And who builds houses? We do, as construction people. And the UK needs more houses, hundreds of thousands of them. So that's one area where you might find yourself employed… in house-building. Then there's infrastructure. We need more dams or reservoirs, that's for sure. Global warming, if that's what you want to call it, means the earth is heating up. We'd better capture that rain when it does fall, because we sure can't live without it.

Here's an idea. Commission five reservoirs across the country. In times of recession, it will keep the country on track, giving work to people to feed their families and to future-proof the country for their material needs. We can't survive without water. And it rains all the time. Why waste it? We need it. Capture it. The hundreds of thousands, or is it even millions of homes that we need to build, they will all need water. Where is it coming from? I read the other day that England will face a water crisis within twenty-five years. Guess what. These reservoirs take years to build. The design, tendering, and commissioning process takes years. Why not begin it now? The build will take years. It's all construction. It's all jobs. And it all involves us as QS's.

Then there are roads, and then there's rail, and then there's schools, hospitals, bridges, and shopping centres, and cinemas and stadiums, and Olympic villages and amusement parks and whatever else you might think we might need in the next hundred or thousand years. And all of this stuff has to be built by someone, and that someone can be you, or me, or our children, and our children's children.

And once it is built, it needs to be maintained. And when it's old, and it hasn't been properly maintained, it needs to be restored; it's all construction, and it all needs a team, and that team needs a manager, and that team needs someone to manage the costs.

That's the Quantity Surveyor.

That's you.

OTHER BOOKS

From the author of
The Quantity Surveyor's Bible

Israel and Palestine: The Complete History [2019 Edition]

Israel and Palestine: The Complete History seeks to explain the overall story of Israeli and Palestinian tensions and divisions in the region. Indeed, without properly understanding the full history of the area, it is impossible to understand the current situation.

In this book, author Ian Carroll takes the reader back to the very beginning of the conflict some 4,000 years ago, then moves through the major events of the Middle Ages and 20th century, and brings us right up to the present day, documenting the significant events that have happened along the way. The reader is allowed to make up their own mind as to where praise and condemnation belong with this complicated issue.

From Exodus to the birth of Jesus, from Islam to the Crusades, through the Diaspora and up to the recreation of the modern state of Israel and beyond, Israel and Palestine: The Complete History avoids a dry academic approach. It aims to tell the history of the region and peoples in a balanced and brisk fashion, from a storyteller's perspective.

Togetherness: How to Build a Winning Team by Dr Matt Slater

Togetherness is a powerful state of connection between individuals that can lead to amazing triumphs. In sport, teams win matches, but teams with togetherness win championships and make history.

If you want the individuals on your team to develop their skills and reach their potential, get them 'together'. The key to this, is to understand your players' group memberships and how to harness them, to create a unique team identity that is special to "us".

This concise and practical book – from Dr. Matt Slater, a world authority on togetherness – shows you how you can develop togetherness in your team. The journey starts with an understanding of what underpins togetherness and how it can drive high performance and well-being simultaneously. It then moves onto practical tips and activities based on the 3R model (Reflect, Represent, Realise) that you can learn and complete with your team to unlock their togetherness.

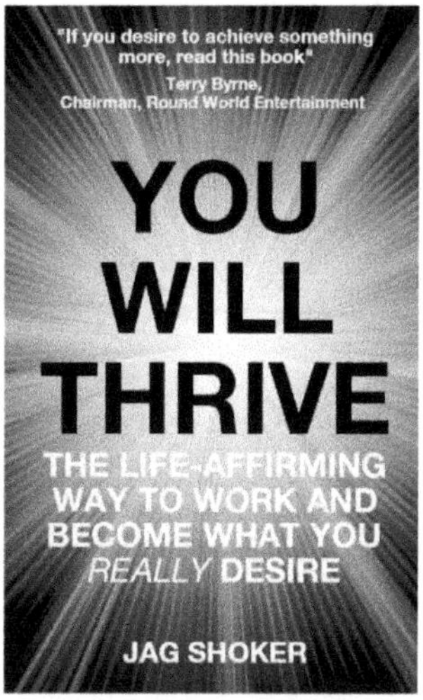

You Will Thrive: The Life-Affirming Way to Work and Become What You Really Desire by Jag Shoker

Have you lost your spark or the passion for what you do? Is your heart no longer in your work or (like so many people) are you simply disillusioned by the frantic race to get ahead in life? Your sense of unease may be getting harder to ignore, and comes from the growing urge to step off the treadmill and pursue a more thrilling *and* meaningful direction in life.

You Will Thrive addresses the subject of modern disillusionment. It is essential reading for people looking to make the most of their talents and be something more in life. Something that matters. Something that makes a difference in the world.

Through six empowering steps, it reveals 'the Way' to boldly follow your heart as it leads you to the perfect opportunities you seek. Through every step, it urges you to put a compelling thought to the test:

You possess the power within you to attract the right people, opportunities, and circumstances that you need to become what you desire.

As you'll discover, if you find the *faith* to act on this power and do the Work required to realise your dream, a testing yet life-affirming path will unfold before you as life *orchestrates* the Way to make it all happen.

The Savvy Traveller Survival Guide by Peter John

Travel is one of our favourite activities. From the hustle of bustle of the mega-cities to sleepy mountain towns to the tranquillity and isolation of tropical islands, we love to get out there and explore the world.

But globe-trotting also comes with its pitfalls. Wherever there are travellers, there are swindlers looking to relieve individuals of their money, possessions and sometimes even more. To avoid such troubles, and to get on with enjoyable and fulfilling trips, people need to get smart. This book shows you how.

The Savvy Traveller Survival Guide offers practical advice on avoiding the scams and hoaxes that can ruin any trip. From no-menu, rigged betting, and scenic taxi tour scams to rental damage, baksheesh, and credit card deceits – this book details scam hotspots, how the scams play out and what you can do to prevent them. The Savvy Traveller Survival Guide will help you develop an awareness and vigilance for high-risk people, activities, and environments.

Forewarned is forearmed!